Praise for
THE FLAGSHIP EXPERIENCE

"Automobiles are an important part of our lives, yet we mostly leave them idle. With connectivity, electrification, and autonomous driving, our cars could always be 'on' like our mobile devices today. Today, we choose our cars based on design, comfort, quality, and the driving experience—areas incumbent automakers understand well. But will the time come when the software experience supersedes the physical experience as the customer's primary consideration, and consequently the software brand identity overshadows the hardware's? The automaker that balances the physical with the digital customer experience will likely prevail. *The Flagship Experience* offers a timely perspective."

NEIL BROOKER, **head of business transformation and foresight, BMW Designworks**

"Software-Defined Vehicles are big data generators. The insights derived from this data can help automakers improve customer loyalty and monetization, and attract new customers. *The Flagship Experience* presents a comprehensive road map and easy-to-understand framework for achieving these goals. It should become an indispensable resource for every automaker and their partners."

KISHOR PATIL, CEO of KPIT

"The Software-Defined Vehicle, enabled by AI, opens new ways for automakers to delight their customers. With the right service and business model, it can create additional revenue opportunities along the lifecycle. *The Flagship Experience* presents an easily digestible framework for the automotive industry to achieve the full potential of its digital technology investments."

JONAS SEYFFERTH, director digital mobility at PwC Strategy&

"Automakers have gotten by for a century delivering incremental technological change on long product cycles, while outsourcing customer relations to dealers. That strategy is doomed. Evangelos Simoudis tells you why, and what legacy automakers must do to survive. Every business leader can learn from this."

JOSEPH WHITE, global automotive correspondent, Thomson Reuters

"Similar to how the electric vehicle revolution became the mobility story of this decade, the Software-Defined Vehicle revolution will be the story of the next decade and beyond. It will get local, state, and federal governments to zero fatalities on their roadways and smart transit solutions to every city—big and small, rich and poor. *The Flagship Experience* eloquently articulates how the present will inspire the future, and how every citizen of humanity can benefit from this new technology."

TREVOR PAWL, former chief mobility officer for the State of Michigan

ALSO BY EVANGELOS SIMOUDIS

The Big Data Opportunity in Our Driverless Future (2017)

Transportation Transformation (2020)

www.amplifypublishing.com

The Flagship Experience: How AI and Software-Defined Vehicles Will Revolutionize the Automotive Customer Experience

For more information, please contact:
Amplify Publishing, an imprint of Amplify Publishing Group
620 Herndon Parkway, Suite 220
Herndon, VA 20170
info@amplifypublishing.com

Library of Congress Control Number: 2023916800
CPSIA Code: PRV0124A
ISBN-13: 979-8-89138-061-5

Printed in the United States

The Flagship Experience

How AI and Software-Defined Vehicles Will Revolutionize the Automotive Customer Experience

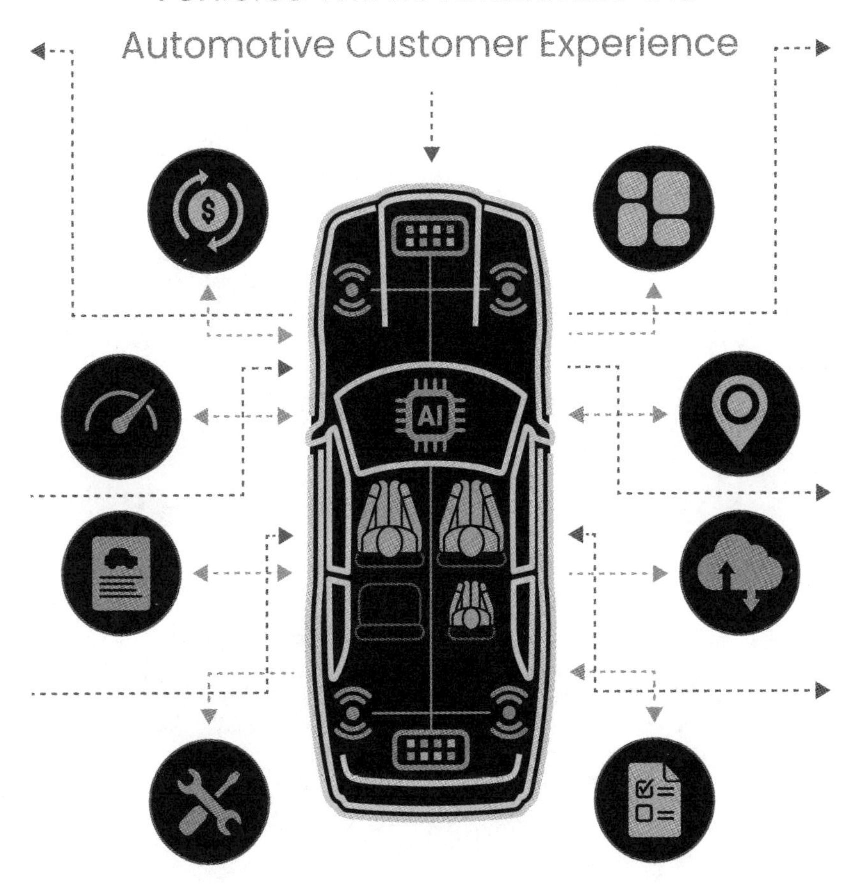

EVANGELOS SIMOUDIS, PhD

Foreword by Timothy Morey

CONTENTS

FOREWORD i

CHAPTER 1
The Need for a New Automotive Customer Experience 1

CHAPTER 2
The New Mobility 15

CHAPTER 3
The Flagship Experience 29

CHAPTER 4
The Software-Defined Vehicle 63

CHAPTER 5
Software for the Software-Defined Vehicle 87

CHAPTER 6
Software for the Flagship Experience 109

CHAPTER 7
Organizational and Business Model Transformations 137

CHAPTER 8
Conclusion 179

REFERENCES 191

ACKNOWLEDGMENTS 209

ABOUT THE AUTHOR 211

INDEX 213

Foreword

In the summer of 2012, I found myself in a meeting with executives from a leading Japanese carmaker at the San Francisco offices of the design consultancy I was working with at the time. The executives were on a tour of Silicon Valley, meeting with technology companies and creative consultancies to explore the potential of using software to compete on customer experience. The carmaker's leaders had recognized that they were in a period of value migration, moving from competing on powerful engines and beautiful design to competing on a more comprehensive mobility experience that integrated with their customers' digital lives.

They realized that their current skills and competencies were not aligned with the future demands of their industry. They almost entirely lacked internal software design and development capabilities. They didn't have a relationship with their end customers because that was handled through dealerships. They didn't have access to their customers' data to understand how the vehicles are used. And even if they had access to their customers' mobility data, they lacked internal data science capabilities to analyze it. Their partner ecosystems primarily consisted of automotive parts suppliers. But above all, their culture and mindset were rooted in manufacturing excellence rather than prioritizing customer experience. A lot has changed in the past decade in the automotive and mobility industries. However, the wholesale transformation of carmakers from being hardware-centric manufacturers to purveyors of software-centric customer experiences is the most consequential and is well underway.

These moments of transition are dangerous for industry incumbents. Consider the switch from feature phones to smartphones—the companies that emerged as victors in the smartphone market were mostly newcomers with a background in computing rather than telecommunications. Today, automotive OEMs face a similar technology transition away from internal combustion engines to electric vehicles, as well as a consumer value proposition transition from driving to a more comprehensive mobility experience. Yet despite these challenges, the automotive OEMs also have tremendous advantages. They have trusted brands that they have nurtured for decades. They have manufacturing expertise to build high-quality cars at scale. They have well-functioning supply chains.

In this book, Evangelos Simoudis charts a path for both incumbents and newcomers in the industry to navigate this new competitive landscape. At the heart of this strategy is the notion that OEMs must find ways to improve post-sale customer monetization. OEMs must build and maintain a relationship with their end customers and find ways to increase their customers' lifetime value. The one-time transaction of selling a new car to a buyer through a dealership entirely misses the opportunity to create additional value for that first buyer, let alone subsequent owners over a vehicle's lifetime.

The ideal way for automotive OEMs to increase their customers' lifetime value is by offering a Flagship Experience, which builds on the idea of the Flagship Car. For the past fifty years or so, carmakers have been selling aspirational brand images. The Flagship Car has been the manifestation of that aspiration for each brand. Today we are moving to an era when carmakers compete on customer experience, hence the Flagship Experience is the new way in which carmakers deliver on their brand promises.

Several technologies contribute to delivering the flagship experience. First is the transition to Software-Defined Vehicles that enable carmakers to update, change, and alter cars throughout the vehicles' lifecycle. Software-Defined Vehicles are vehicles that are based on alternative powertrains, primarily battery electric, incorporate a level of driving automation that provides them with enhanced safety characteristics, and have novel architecture that enables their easy configuration and continuous updating. Rather than being fixed the moment the car rolls off the production line, the car receives over-the-air updates and continues to evolve. The underlying Software-Defined Vehicle architecture can be tailored to deliver flagship experiences across a carmaker's entire fleet, and not just at the premium end of the market. Second, the rapid advancement in AI allows automakers to better understand their customers and tailor experiences to meet their needs. The promise of personalized customer experience has been with us for many years; AI delivers on the promise.

As the competitive landscape evolves from making beautiful cars with powerful engines to designing delightful mobility experiences built on Software-Defined Vehicles, the canvas within which to innovate becomes much broader than before. Carmakers are given license to create value for their customers well beyond simply making great cars. By understanding the mobility-related goals customers aim to achieve—the "jobs to be done"—carmakers can offer innovative services, products, and experiences that go far beyond their current offerings. This is, of course, challenging for incumbent carmakers, but it also opens a lot of new opportunities.

One of the highlights of Evangelos Simoudis's thinking and writing lies in his practical frameworks. I have personally used frameworks from his previous books, *The Big Data Opportunity in Our Driverless Future* and *Transportation Transformation: How Autonomous Mobility Will Fuel New Value Chains,* in various automotive and mobility projects. These frameworks bring clarity and simplicity to a complex and ever-evolving ecosystem. Evangelos Simoudis combines the logic and analytical rigor of an engineer with the vision and imagination of a futurist and investor, bringing order to what might otherwise seem like an overwhelming and chaotic domain.

Whether you are an entrepreneur or investor looking for opportunities in the emerging mobility marketplace or an incumbent carmaker working to evolve your organization to succeed in the future, this book has valuable insights for you. It provides an excellent overview of competing on customer experience throughout the mobility journey, while also offering detailed bulleted lists and actionable frameworks to guide your thinking on building the new flagship experience with Software-Defined Vehicles. We are embarking on a journey where the destination is not entirely clear. There will be strategic twists and turns, technological advancements, and undoubtedly some setbacks along the way. However, this book illuminates the path forward and equips you with the clarity needed to take those initial steps. I am excited to witness the advancements we will make in mobility in the coming years.

Timothy Morey
Global Head of Method,
a Global Logic Company

CHAPTER 1

The Need for a New Automotive Customer Experience

The automotive industry has entered a period of radical transformation. Automakers that fully participate will become customer-centric corporations, change the way they conduct business, and tap new revenue streams. During the 135 years of its existence, the industry has experienced two transformations of similar magnitude,[1] both of which were vehicle-centric. The key drivers of this transformation are the transition from fossil fuels to renewable energy to address climate change, our changing mobility patterns, and disruptive technology innovations in batteries, semiconductors, software, and AI. The context for this transformation is provided by "new mobility," a global movement to redefine how we transport people and goods. New mobility changes the rules for the automotive industry because it brings together social awareness about the environmental impact of

1 The invention of the assembly line led to the first; the Toyota Production System sparked the second (Ohno & Bodek, 2021).

transportation modalities, vehicles based on new architectures and powertrains, digital platforms that manage transportation, and new business models to monetize it. The enduring point of connection between these elements—the mobile unit in new mobility—is not the vehicle, but the customer. As they attempt to become customer-centric companies, automakers must rethink not only the vehicles they introduce but the experience and services they will offer around them.

Consumers and companies alike are starting to consider seriously the need for transportation to transition away from fossil fuels to combat climate change. The drivers of vehicles using alternative fuel powertrains, the passengers of autonomous ride-sharing vehicles, the users of electric bicycles and scooters, or even the operators of goods delivery fleets are increasingly concerned about their environmental impact. As a result, they are adopting new work patterns, embracing new approaches to urban living, and having an active digital life that minimizes the need to move.

Automakers are responding to customer demand for vehicles that are based on alternative fuel powertrains by introducing electric vehicles. In addition to new battery and electric motor technologies, these vehicles, called Software-Defined Vehicles, are based on new architectures and incorporate disruptive semiconductor, software, and AI technologies. They are software-centric, compute-intensive, and intelligent. The multitude of sensors with which they are equipped make them big data generators. As we will see later in the book, their architectures enable their extensive on-the-fly reconfiguration.

To bring Software-Defined Vehicles to market, automakers are making large investments to build new manufacturing facilities, establish new supply chains, and recruit personnel with skillsets not traditionally associated with the automotive industry. They have committed to investing over $1 trillion by 2030 for the development of electric vehicles (Lienert, 2022). Many of these capital infusions are the result of partnerships, such as GM's partnership with Honda to build affordable compact electric vehicles (Capparella, 2022). Incumbents are also starting to realize that these new vehicles will require new rules/new processes for design, manufacturing, and support. But if they remain exclusively focused on designing and manufacturing vehicles, they will be missing the big opportunity to understand from the ground up the new mobility behaviors their customers are exhibiting and take advantage of them by improving customer loyalty and monetizing the customer on a recurring basis based on the customer lifetime value. Customer lifetime value is the present value of the business attributed to the customer during the entire life of his relationship with the company. The automotive industry produces annually close to 80 million new light-duty vehicles worldwide. In most cases, each of these vehicles is used by more than the person who acquires it. The industry must find a way to monetize every user of each vehicle. To achieve these goals, the OEM needs to own the customer relationship and provide a new customer experience. The technologies incorporated in the Software-Defined Vehicles and the new mobility platforms will enable OEMs to understand how, when, and why their customers move, regardless of whether they are using their personally owned vehicle or an alternative modality. Through the customer experience, the OEM will capture the generated data, exploit it using advanced AI techniques, generate the right value propositions, and make them available to the customer at the right time through the vehicle's software or other channels available to the automaker.

Motivated by his concern about climate change, a friend recently bought a foreign automaker's top-of-the-line electric vehicle model. He also added solar panels to his suburban house so that he could charge the vehicle while at home. I met with him after he hadtaken a long-weekend road trip with his family, which prompted a vow never to take another similar trip with the vehicle unless he can be certain that he will not have to charge it on the road. He hadn't identified the right charging stops in advance. In the US no single mobile application provides comprehensive information about charging stations. Finding a functioning charger in a place that felt safe for him and his family to park and wait for the vehicle to charge proved much harder than he had imagined. As he put it, in several places the chargers were not working, or they were slow, or were in places where he didn't feel safe to stop. I told him that he was fortunate in having a house where he could install solar panels and charge the vehicle. Few households around the world have access to this convenience. Most electric vehicle owners live in apartments and need to recharge their vehicles using public chargers. It is important to understand the context created whereby the owner of the electric vehicle, and by extension the Software-Defined Electric Vehicle, operates because the context impacts the customer experience that is expected. To a consumer, long-distance travel using an electric vehicle creates a different context for charging than everyday city travel. And everyday city travel creates a different context if the consumer owns their house versus living in an apartment. To a family taking a long-distance trip, the electric vehicle defines a completely different context; the vehicle is a place of social interaction and entertainment. To a business operating a fleet of electric vehicles, urban driving defines one context: the vehicle as a workplace. Charging the urban fleet can be addressed by a central charging station that can accommodate all vehicles in the fleet. Charging the long-distance travel fleet requires the existence of a distributed network of charging stations that operate reliably. The customer experience needs to view the new vehicle as a transportation modality, a space for the driver and the passengers to interact and even be entertained, and as a workspace and object that operates in a larger environment, e.g., a fleet or a city's transportation infrastructure.

With conventional Internal Combustion Engine (ICE) vehicles, we don't see refueling as part of the customer experience for which the automaker is responsible. If we don't like something about a gas station we've stopped at—its prices, its location, even its vibe—we know that we can drive a little farther and find a more agreeable alternative. We may also trust a particular gasoline brand and see each of its stations as an extension of the brand. This is not the case with charging stations and the companies that own and manage them. At least not yet. Today most of us do, however, consider electric vehicle charging part of the automaker's customer experience. Tesla made sure of that. The importance of charging in the automotive customer experience drove the recent decision by GM, Ford, and Rivian to adopt Tesla's charging adapter standard, implicitly admitting that Tesla is ahead of them in charging technology and station quality and availability (Ewing, 2023).

Our changing views on mobility and the activities that are part of the automotive customer experience represent a challenge and an opportunity for the automaker. It is a challenge because the automaker must now address factors that it didn't have to before. For example, a greater variety of items are delivered to us now, e.g., groceries, merchandise, and restaurant orders, than three years ago. With the Software-Defined Vehicles themselves, in addition to vehicle safety, fuel efficiency, vehicle quality, price, and total cost of ownership, the automaker must now address new areas such as electric vehicle charging, training the driver on how to use the increasingly automated vehicle, and online vehicle configuration and preordering.

This is an opportunity because by analyzing the generated data, the automaker can learn, in a privacy-preserving way, the customer's mobility behaviors inside and outside their privately owned vehicle. By combining the understanding of these behaviors with the customer's lifetime value, the OEM can determine how much "pampering" the customer warrants and how much it can afford to "invest" in each customer relationship. Consider two examples. By understanding how frequently and for what purpose the customer uses their vehicle, the distances traveled, and their lifetime value, the OEM can offer the customer an insurance policy that is priced below the one the customer is getting from their current insurance carrier. Maybe even offer the first three months for free if the customer agrees to sign a two-year contract. Such offers enable the OEM to establish a richer customer relationship that goes beyond the vehicle acquisition transaction. Moreover, this type of relationship is monetizable on a recurring basis. By informing the customer when and where to charge their Software-Defined Vehicle, the OEM uses recharging to provide the customer with immediate value. Providing value continuously is the foundation of a deeper, enduring, and ultimately profitable customer relationship.

As we will see later in the book, the Software-Defined Vehicle is the only way for the OEM to collect rich data about the customer's context-specific mobility behavior and the vehicle's performance and easily apply the knowledge generated from this data using AI in ways that provide customer value. Conventional connected vehicles today generate data, though not as much as Software-Defined Vehicles do, but applying the knowledge extracted from this data back to the vehicle is tedious and at times impossible. In addition to contributing to the customer relationship, predictive knowledge about the vehicle can help the automaker improve its software, e.g., the software managing the vehicle's battery performance. Both personalization and performance improvement positively impact the customer experience.

Since this was my friend's first electric vehicle, I inquired further about the acquisition phase of his customer journey. Surprisingly, he focused not on the dealer experience but on three fragmented customer experiences: buying the vehicle (pleasant as it was); acquiring and installing the solar panels, home batteries, and vehicle charger; and registering for the entitled government rebate. My friend had to learn separately and on his own about the home equipment and the government process. A short time ago, nobody would have thought of these disparate experiences as being connected and part of the automotive customer experience. Yet they are. Bringing them together is another new opportunity for automakers, particularly those targeting more affluent customers, as they value service quality most highly. My friend's vehicle acquisition journey and post-acquisition charging station experience, including his range anxiety and concerns about charging time and station safety, are just a few of the issues that automakers will need to address in transforming their customer experience.

This book presents a blueprint for a new customer experience to be offered by automakers around their Software-Defined Vehicles and provides initial guidelines for its implementation. This customer experience, which I call the Flagship Experience, has customer-centric and product-centric characteristics. It is customer-centric in that it covers the *end-to-end* relationship between each customer and the automaker. This relationship extends outside the vehicle's cabin and lasts over all the Software-Defined Vehicles the customer acquires during its course. The Flagship Experience can also be simultaneously considered product-centric in that it is offered by the automaker to *every* owner of each Software-Defined Vehicle and not just the first. Think of it as creating a "vehicle journey" that tracks the vehicle throughout its life. It enables the OEM to reconfigure and update the Software-Defined Vehicle before making it available for resale and enables each new owner to modify the used

Software-Defined Vehicle in the ways they desire. For the Flagship Experience to succeed in its dual role, the OEM must own the customer relationship and view its Parc[2] as a fleet it can control.

To flourish in the era of new mobility, automakers must shift from being vehicle-centric to becoming customer-centric and offer a version of the Flagship Experience with every Software-Defined Vehicle in their lineup. This combination of vehicle and customer experience and controlling the Parc will strengthen their customers' loyalty and increase the monetization of every customer and vehicle. OEMs are already investing heavily to develop Software-Defined Vehicles and are committing to continue their investment pace. But in today's competitive environment, which includes newcome disruptors, Chinese OEMs, and increasing use of mobility-related services, incumbents will not be able to generate positive returns on their invested capital solely through the sale of Software-Defined Vehicles. They will need to monetize their business and consumer customers through services offered based on a granular understanding of each customer's value. This understanding will come from the analytical engine at the heart of the Flagship Experience: the use of AI to exploit the data generated by Software-Defined Vehicles and data about consumer mobility behavior.

For the Flagship Experience to succeed, the automakers must own the customer relationship and convince the customers to share their data and acquire services from them instead of, or in addition to, technology companies, including Google, Apple, Amazon, and others, as they are doing today. OEMs must become the best in understanding the customer's mobility-related contexts and needs. They must offer services under value propositions that the customer will find difficult to turn down, even if competing

2 Parc is a European term for all registered vehicles within a defined geographic region, originating from the French phrase *parc de véhicules*, meaning the collective number of vehicles or a vehicle collection.

services are available and services that their competitors, including the nonautomotive ones, cannot offer. If neither of these can be achieved, then the companies that today own the digital life experience will be in the position to capture the lion's share of the customer's mobility-related monetization.

To successfully implement and deploy the Flagship Experience, automakers will also need to hire the right people and adopt a culture of agility and experimentation in which to develop services and business models. These will be implemented on early adopters of Software-Defined Electric Vehicles, a market segment that tolerates the testing of new ideas, from new features to new processes and business models. Experimentation, combined with extensive data collection and in-depth analysis, will enable automakers to determine which features and services are embraced, which can be monetized using specific business models, and which will be offered at no extra cost. Agility will enable automakers to easily incorporate within their Flagship Experience the features and services customers value and need for each vehicle class, as well as devising new experiments to provide them with a deeper understanding of their customers.

The choice of business model will present incumbent automakers with a particularly difficult challenge because, in addition to revenue and margins, it will impact the customers' willingness to acquire a service and even their loyalty. In the interviews I conducted with automotive executives during my research, I could only get high-level comments about the results their companies have achieved to date from the business models they tested with the services they offered in conjunction with their connected vehicles. Automakers must consider business models that worked successfully, such as subscriptions to popular consumer services like Amazon Prime, modifications to business models that stopped being effective, such as Netflix's introduction of advertising models to supplement the company's subscription model, and others that didn't work as planned, such as Uber's or Lyft's unprofitable ride-hailing transaction model.

Incumbent automakers will need to remake their partner ecosystem, starting with their dealers. Partnerships will be defined around the OEM controlling the customer relationship and the partner providing continuous value to that relationship. Today's dealers will be able to provide value in the new ecosystem that will emerge but under different conditions. For example, dealers can be used to educate prospective customers on the Software-Defined Vehicle's characteristics to help them with vehicle configuration and operation. Consider a configuration-related issue like the vehicle's battery size. The battery cost is a major contributor to the price of these vehicles. On the one hand, customers want vehicles with large batteries to address their range anxiety, but on the other hand, they should understand that large batteries take longer to charge, resulting in heavy vehicles which could damage roads, and in some instances because of the energy they require, may be more harmful to the environment than some gasoline-powered vehicles. We are already starting to see how such partnerships may work through examples from the car rental industry. The car rental company Hertz, GM, Tesla, Polestar, and the city of Denver have recently launched a program to educate drivers, including ride-hailing drivers, on the characteristics of electric vehicles, which includes information on adjustments the driver will have to make to facilitate the charging process and avoid range anxiety or other inconveniences (Muller, 2023).

Radical transformation is not easy for incumbent automakers. Software-Defined Vehicles and the Flagship Experience will require significant investments for several years. Incumbent automakers must commit to making these investments even though today their ability is being questioned. The demand for Software-

Defined Vehicles is inconsistent and softens every time government subsidies are eliminated. These vehicles will remain unprofitable for several more years due to their high development costs. In the meantime, competition will remain fierce and negatively impact profit margins. The competition will come from newcomer automakers that are already offering comprehensive customer experiences with their Software-Defined Vehicles and Chinese automakers that are increasingly entering the global market with lower-priced Software-Defined Vehicles boasting impressive features. Incumbent automakers will also compete for mobility-related services revenue with the technology companies that own major components of the consumer digital life and are aggressively entering the automotive industry. As the sales of ICE vehicles slow down due to economic and geopolitical reasons, the profits from such sales that have been used by incumbents to partly fund the development of their Software-Defined Vehicles will no longer provide a reliable funding source.

Each automaker will choose its road ahead. Most are already transforming to offer Software-Defined Vehicles with a conventional customer experience. But with so many changes impacting mobility, it will be important to break from how the automotive customer experience has been approached conventionally and consider an end-to-end mobility experience. Those who adopt the approach recommended in this book will transform in a way that enables them to achieve the promise offered by wrapping each Software-Defined Vehicle with the right Flagship Experience and achieve the promise of the automotive industry's third radical transformation.

Chapter 2, The New Mobility, introduces new mobility, describes the three phases over which it will evolve and the mobility patterns emerging during each phase, and outlines the issues that will need to be addressed as automakers prepare to offer Software-Defined Vehicles.

Chapter 3, The Flagship Experience, defines the Flagship Experience, contrasts it to today's automotive customer experiences, demonstrates how it can be tailored to each customer's needs, and explains how it can be monetized by the automaker and its ecosystem partners.

Chapter 4, The Software-Defined Vehicle, describes the architectures and platforms that comprise the Software-Defined Vehicle and explains the business implications of the automakers' architecture choices in offering the Flagship Experience.

Chapter 5, Software Technology for the Flagship Experience, focuses on the software development methodologies that must be adopted to successfully implement the Software-Defined Vehicle's architecture, the software tooling that must be employed, and the technology-related transformations that must be undertaken before an automaker can roll out the Software-Defined Vehicles that will be able to fully exploit the Flagship Experience.

Chapter 6, Customer Analytics Using the FEAT Framework, introduces the FEAT customer understanding AI framework and demonstrates how to use the analysis results to deploy the Flagship Experience so as to lead to customer monetization. Monetizing the Flagship Experience and achieving the revenue goals projected by automakers will depend on the characteristics of the segments that adopt it.

Chapter 7, Organizational and Business Transformations, presents the past efforts by OEMs to implement organizational and business model transformations, analyzes the organizational changes that OEMs have recently undertaken in their effort to address the industry's radical transformation, and offers recommendations on how to address certain challenges depending on the sequence of the selected transformations.

Chapter 8, Conclusions, synthesizes the conclusions as to how the Flagship Experience will help automakers compete effectively with customers and achieve the financial goals they associate with their Software-Defined Vehicles.

CHAPTER 2

The New Mobility

Following almost three years of restrictions due to the pandemic, mobility is recovering. But it is different. Our movement decisions today are impacted by our concerns about climate change (shouldn't I look for a zero-emissions vehicle?), our adoption of hybrid work (why travel to the office if I can accomplish my goal via a video conference?), and our continued embrace of many of the conveniences adopted during the pandemic (why not have my groceries delivered instead of going to the grocery store?). Even though the personally owned vehicle remains the primary transportation modality, automakers must understand how the answers to these questions impact their customers' mobility, the choices customers make about the vehicles they acquire, and the way they use them. Their understanding and interpretation will impact the customer experience they choose to offer around their Software-Defined Vehicles.

Notice that I wrote "mobility" and not "transportation." We often use the two words interchangeably, but they are not synonymous. Transportation refers to moving people. Mobility refers to the ability to move or be moved. Mobility is about having transportation options, such as a personally owned vehicle or reliable public transportation, and relying on them to take you where you need to go or bring you what you need to get. It involves understanding the characteristics of those options in terms of environmental impact, safety, affordability, convenience (including travel time, charging infrastructure availability, charging time, etc.), and the overall traveler experience they provide. Should I drive my electric vehicle to my destination today or use a combination of other modalities such as train and ride-hailing? If I drive my electric vehicle, will I be able to find a parking garage with a charging station near my destination? How expensive will it be? Will the charger be working? Will it be in a safe area given the time that my meeting will likely end? These are questions that consciously or unconsciously we try to answer every time we are about to have a transportation experience, depending on the mobility options available to us.

2.1 The Changing Mobility Patterns

New mobility patterns are emerging from relocation, whether by choice or due to employment, changes to how we approach commuting and shopping, and concerns about the environment. To develop the Flagship Experience and tailor it to each customer, automakers must understand such patterns because they impact vehicle purchase activity, the types of vehicles consumers and businesses acquire, and the expectations during the customer journey.

Urban populations around the world are growing. Based on the most recently published UN reports, 3 million people a week around the world are moving to cities (Boyd, 2021). But the pandemic also led to urban population shifts from large cities to smaller, more affordable ones, and from city centers to the suburbs, where space made working from home more manageable (Frey, 2021; Smith D, 2020; Yu and Melgar, 2021). Because neither small cities nor suburbs provide good transportation options, we end up using privately owned vehicles. We now see that smaller cities with significant population growth during the pandemic face traffic congestion problems (Stewart, Cotret, & Davidson, 2021). Moving to the suburbs results in urban sprawl, which also impacts mobility because it makes certain trips longer. While such population shifts may be more pronounced in the US, where changing places of residence is not uncommon, during the pandemic we also saw it in the EU, particularly from the western and northwestern countries of the Union to the southern and eastern members, with people wanting to return closer to their families (*The Economist*, 2021).

In many cities, the pandemic has changed the frequency and the way urban dwellers move daily (Dizikes, 2023). People living in these cities are a) adopting zero-emission vehicles faster than ever before, particularly in Asia and Europe, b) moving less because items or services are delivered to them, c) traveling shorter distances that in many cases can be covered either by walking or using micromobility, d) relying less on public transportation, preferring instead to use multiple modalities in the same trip, and e) socializing differently (Cantor, Soulopoulos, Fisher, O'Donovan, and Cheung, 2022; Irle, 2022; Wilson, 2022; North American Bikeshare and Scootershare Association, 2022; Arriva, 2021).

Factory automation is accelerating and is impacting blue-collar labor forces (Casselman, 2021). Because of the introduced automation, companies are employing fewer workers per shift. In the case of automotive manufacturing, we are also seeing fewer shifts (some of which are attributed to lower demand) and sometimes shorter shifts to preserve employment. These changes are also impacting mobility patterns. In some cases, we may see factory workers traveling less, or traveling during different times of the day compared to what had been established norms, especially in factory cities like Detroit, Stuttgart, and others. In other cases, we may see them traveling long distances or having to relocate to places where they could find work.

Commuting to and from work, normally about 20 percent of our weekly trips in the US, has decreased significantly, especially in places with a high concentration of service and knowledge workers. Many companies have adopted telework, hybrid work, and flexible work hours for white-collar workers. The adoption of telework and hybrid work by more companies from larger industries will continue to lower the number of commuting-related trips. Telework doesn't have to be done necessarily from home, but from facilities that are closer to the home, resulting in shorter trips. Many white-collar workers used these changes to either become digital nomads or permanently relocate to different areas (Lufkin, 2021; Bowman, 2022).

Shopping trips, normally 30 percent of the weekly consumer trips in the US, have decreased significantly because of the accelerated growth in e-commerce (Brewster, 2022). The adoption of e-commerce by more population segments globally will continue increasing the number of goods delivery trips (Bennett, 2023). On-demand goods delivery could become a more important and profitable business than passenger transportation.

E-commerce and goods delivery were clear winners during the pandemic. While the growth rates of companies in these two areas have moderated in the post-pandemic period, they remain higher than the pre-pandemic rates. Companies that offer on-demand goods delivery, including grocery delivery, and restaurant delivery continue to grow, even as companies like Walmart, Target, and others offer their private delivery services. People grew accustomed to ordering from home and in the process also became used to cashless and contactless commerce. This practice provides convenience because it reduces personal trips to supermarkets, big-box retailers, etc., and is more environmentally friendly because each delivery vehicle fulfills several deliveries during a trip. Terrestrial and aerial autonomous vehicles used for short- and long-haul goods delivery will extend the opportunity even further (Duffy, 2020).

2.2 The Three Phases of New Mobility

Transportation planners and academics created the field of new mobility to address transportation's environmental impact, increase its accessibility to larger parts of the population, and make it safer. New mobility addresses emerging movement patterns but also leads to new patterns. I define **new mobility** as the ability to move people and goods in environmentally friendly ways, safely, affordably, and conveniently, using a coordinated combination of intelligent zero-emission vehicles and shared transport services that are offered on a scheduled or as-needed/on-demand basis.

OEMs must understand new mobility's implications for their business. Their Software-Defined Vehicles will need to operate within regional new mobility contexts. As we will see below, as it advances new mobility will be increasingly multimodal and services-based. The companies providing these services will use fleets of Software-Defined Vehicles. This means that the services OEMs offer to such fleets could provide them with a large and important revenue stream. GM's BrightDrop and Ford's Ford Pro will be big beneficiaries of this trend, and the financial results these businesses already report provide an early sign of their potential. But the move to multimodal and services-based mobility also impacts the Flagship Experience narrative. *Why, how, where,* and *when* we move are all questions at the core of new mobility and impact the customer experience of both consumers and fleet operators, making the Flagship Experience relevant to both constituencies. Answering a question such as "What modalities does the traveler use to get to the intended destination?" relates to the transportation's environmental impact, safety, affordability, and convenience, while answering "What is the best way to deliver packages to specific destinations to optimize the electric van's range?" impacts a fleet's economics.

Cities globally that are adopting new mobility are investing in public transportation and transportation infrastructure, restricting the use of private vehicles in certain areas while encouraging the use of other transportation modalities, e.g., micromobility, but also requiring higher population densities in certain areas (Berg, 2022; Peters, 2020; Kober, 2022). Intelligent and highly instrumented physical and digital transportation infrastructures, such as those deployed in Singapore and Dubai, enable the *coordination* of privately owned vehicles and vehicles used by fixed-schedule and on-demand services that move people or goods. The results of these investments can be leveraged by Software-Defined Vehicles and the Flagship Experience. Consider, for example, intelligent journey planners like Moovit's that uses data generated by public transportation systems to help consumers plan their trips within a city. Or Project44's logistics digital platform that takes advantage of the data generated by the public transportation infrastructures to provide supply chain visibility to commercial customers.

I organized new mobility in three phases. Each phase is expected to last several years. During each phase, progress is made toward new mobility's environmental impact, safety, affordability, and convenience. Mobility patterns change as a result of this progress. Understanding the characteristics of each phase provides important cues on the features and services that will comprise the Flagship Experience.

The first phase, the **mobility services emergence** phase, started in the middle of the previous decade. We call it that because it was dominated by the rise of on-demand mobility services for passenger transportation such as ride-hailing and micromobility and goods delivery. The customer experience innovation was the access to such services through mobile applications. Urbanites globally were attracted by the convenience and differentiated customer experience they offered. The use of ride-hailing, vehicle-sharing, and micromobility increased. At that point, mobility entered our digital lives. Automakers tried offering their own on-demand mobility services. Their efforts were unsuccessful and were abandoned. Through their technology innovations, mobility services companies revealed the importance of AI-based digital platforms, e.g., Uber's customer management platform or Cruise's robotaxi operating platform, in understanding and addressing their customers' new mobility needs, as well as in managing and optimizing their corporate operations. For example, Uber's customer-facing application that is part of their customer management platform has evolved from exclusively accessing ride-hailing and ridesharing to supporting other transportation modalities to ordering food and other goods. Uber's loyalty program and membership program (Uber One), which are components of its customer management platform and are accepted by a growing set of partners, encourage the consumer to expand their use of Uber's services. In the US and China, such platforms have been developed by technology companies, whereas in Europe mostly by public transportation organizations. Also during this phase, the transaction-based business models used by the mobility services companies helped us understand this model's deficiencies for monetizing such services.

Companies including automakers, Tier 1 suppliers, technology companies, and startups, developed and started testing automated and autonomous vehicles for ride-hailing and long-haul, middle- and first/last-mile logistics. The efforts led to advances in ADAS technologies that are incorporated into the Software-Defined

Vehicles being introduced. As cities became acquainted with new mobility's advantages and the required investments, during this phase they started considering new infrastructures, fleet electrification, new vehicle technologies to increase safety and convenience, business models to monetize existing and future infrastructures, and new policies to address new mobility's goals.

We have recently entered the second phase, which is the **fleet formation** phase. Capitalizing on the lessons that on-demand ride-hailing companies learned during new mobility's first phase and the introduction of Software-Defined Vehicles that are capable of generating and communicating a variety of mobility-related data about the vehicle and the cabin's occupants, as well as receiving Over-the-Air (OTA) software updates, during this phase automakers and mobility-services companies will be able to manage even the privately owned vehicles as fleets. In the case of automakers, this means that they will be able to treat any part of their Parc they choose as a fleet. Operating a fleet typically results in higher efficiency and productivity and better economics. For example, by organizing and managing all US-based 2023 Mustang Mach-E vehicles owned by consumers as a fleet, Ford can efficiently update their software as a single operation. Similarly, by managing its Tesla Model Y vehicles as a fleet, ride-hailing company Revel can easily forecast their daily charging needs and optimize its charging costs.

Also during this phase, cities partner with mobility services companies to form integrated fleets consisting of public transportation and private vehicles in their continued effort to increase shared mobility and reduce their carbon footprint. Examples of already-formed partnerships include Via's with the public transportation authorities in London and Philadelphia, Grab's partnership with Singapore, Uber's partnership with Atlanta, Berlin's work with a variety of mobility services providers, and Voi's collaboration with the city of Stockholm. Combined with increasingly intelligent journey planning applications, these partnerships will lead to the emergence of new urban mobility patterns as consumers find convenient ways to use multimodal transportation. As they continue to refine their business model and capitalize on the growing demand for goods delivery, ride-hailing companies, including those offering robotaxi service, already started to offer both passenger transportation and goods delivery, and this trend will expand. Combining these functions also impacts the mobility patterns that develop.

OEMs must take four actions during this phase of new mobility. First, they must own the customer relationship. This will allow them to understand and respond to their customers' needs and make it easier to capture a larger percentage of the customer's lifetime value than today when the relationship is owned primarily by the dealer. Second, they must get access to their customers' mobility-related data that does not involve their privately owned vehicles. This will require automakers to partner with cities and mobility services companies that capture this data today. As we will describe later in the book, they will need to combine this data with the data generated in and by their Software-Defined Vehicles and analyze it using AI to tailor the Flagship Experience to each customer. Third, they must control their Parc and offer the vehicles under novel business models, e.g., subscriptions, and manage each vehicle's journey, from the time the vehicle exits the assembly line to the time it is retired. This will allow them to capture each Software-Defined Vehicle's configuration, most of which will be based on software services, and use AI-based vehicle management systems to determine the changes that could be recommended to each vehicle during the ownership period, as well as how the vehicle should be reconfigured based on market understanding before becoming available for a new owner to increase its marketability.

Finally, they must learn lessons from the customer-understanding platforms built by the mobility services companies because they will be required to develop such platforms themselves as part of offering the Flagship Experience.

During the third and final phase, the **mobility-as-a-service** phase, all aspects of the movement of people and goods will be offered as a service. Combined with the continued population movement to cities, mobility-as-a-service could create a significant challenge to automakers as more consumers forego private vehicle ownership. The slowing demand for private vehicles may be partially balanced by the increased demand for vehicles by fleet operators. We do not expect to enter this phase until the middle of the next decade at the earliest.

A few cities will choose to become mobility orchestrators during this phase as they continue their evolution from the second phase of new mobility, e.g., Singapore, out of necessity from such pressures as the rise of sea level due to climate change, or because they are built as such from the ground up, e.g., NEOM in Saudi Arabia (Smith H. , 2021). Orchestrator cities will use advanced AI capabilities to manage urban mobility networks that will offer their citizens a variety of mobility services including automatically determining when a consumer needs to be transported, and with which modalities, and when a good needs to come to the consumer. The transition to this phase will require automakers to think more strategically about the architectures and capabilities of their future Software-Defined Vehicles, and what type of customer experience they will need to offer around them.

The three phases will not advance uniformly around the world, a fact that global automakers need to consider strategically. Phase advancement will depend on geography (economic, political, demographic, the quality and breadth of existing infrastructures, etc.), aligning the different speeds with which cities, mobility services companies, automakers, and customers adopt new mobility and the way each region will choose to address existing and emerging challenges. Most cities around the world will remain simple transportation infrastructure providers. In other words, success with new mobility and the form of this success will be a regional affair often differing from city to city. However, there is no question that public/private partnerships will be key to creating the future we want.

Nobody is expecting privately owned vehicles to disappear even when the third phase of new mobility reaches maturity, but automakers need to understand and appreciate the changes new mobility will bring by the second and third phases because they will impact the vehicle design and ownership experience their customers will respond to. Put another way, the ways new mobility is implemented in a region will impact the types of privately owned vehicles that succeed there. Every industry analyst predicts that the demand for privately owned vehicles will remain healthy for the foreseeable future, and a large percentage of this demand will be addressed by Software-Defined Vehicles using alternative fuel powertrains. Over the next twenty years, the percentage of these vehicles will grow from 5 to 8 percent today (depending on geography) to over 85 percent of all vehicles sold. Moreover, by 2024 at least 50 percent of the new vehicles sold will incorporate at least level two driving automation, reaching almost 100 percent by 2040 (Grant, 2022).

2.3 The Way Forward

During the new mobility's initial phase, OEMs tried various approaches to become relevant in the ensuing transportation transformation (Simoudis, 2020). They started autonomous vehicle programs, offered mobility services, and even ventured into micromobility and microtransit. With these efforts since abandoned, GM's Cruise robotaxi unit being a rare exception, Software-Defined Vehicles that are equipped with driving automation and intelligent cabins represent the automakers' current attempt to address the industry's radical transformation. In Chapter 7 we present the actions they take to address and take advantage of this transformation. However, in addition to investing heavily to develop, manufacture, and support these vehicles, they must start to acknowledge that their future annual sales could be lower for three reasons. First, Software-Defined Vehicles will last longer and be upgradable, giving the customer less of a reason to change them as frequently as today's vehicles. Second, they will not be able to achieve strong sales in regions with reliable new mobility alternatives and low private vehicle ownership levels. Third, they will face stronger competition from local OEMs in previously promising markets such as China.

The new mobility user wants environmentally responsible, convenient, affordable, and safer mobility. Automakers have the potential to play a pivotal role in achieving these goals. Because of how we project new mobility to evolve, taking a vehicle-centric strategy around Software-Defined Vehicles will not be sufficient. Automakers will need to become customer-centric and intimately understand their business and consumer customer needs and monetize them using vehicle-, driver-, and passenger-centric services. These services must support the *appropriate experience*, at the right time, at every opportunity involving mobility. This experience will combine services offered within the vehicle, such as mapping-related content from the vehicle to the cabin's occupants, usage-based insurance, or vehicle anti-theft protection; and outside the vehicle, such as dining locations that match the requester's preferences while visiting a new city during a vacation trip. Such services will need to be offered using business models that today we typically see in technology, e-commerce, and entertainment companies. The intelligence of the Software-Defined Vehicles and AI-based customer and vehicle management platforms will enable the OEMs to provide their customers with *context-specific experiences* in every Software-Defined Vehicle in their lineup, rather than only the top-of-the-line vehicle, as is the case today. Anticipating the traveler's needs and being able to address them at the right time with the right offering will define the Flagship Experience and will be at the core of recurring and scalable revenue models, resulting in enduring loyalty. It also has the potential to result in enduring resilience for the incumbent OEMs overall as they become agile in a changing mobility marketplace, able to anticipate and respond to changing consumer patterns over time.

CHAPTER 3

The Flagship Experience

For about sixty years, well into the nineties, the *automotive customer journey,* i.e., a customer's actions before and after acquiring a vehicle, was simple and linear, with well-defined participant roles and clear handoffs. The automaker was responsible for generating *awareness* and favorable opinions about the model lineup through advertising channels, auto shows, and its dealer network. The prospective customer was a passive participant in the awareness stage. The dealer was responsible for *consideration, intent to purchase, purchase,* and *service,* i.e., helping the customer test each candidate vehicle, convincing him to acquire one of the vehicles available in the lot, working out financing options, servicing the vehicle, and selling parts. The dealer was the interface to the customer and owned the customer relationship, while the OEM remained—and remains, with few exceptions—at arm's length from the customer. Even though the customer journey has become more interactive and certain of its components have been digitalized, incumbent OEMs today remain vehicle-centric. Tesla and other newcomer automakers are customer-centric and are demonstrating that owning the customer and offering a customer experience that is combined with their Software-Defined Vehicle and is integrated with their customers' digital lives gives them advantages that they exploit during every phase of the customer journey and every phase of the vehicle journey, from design to manufacturing to sale, support, and resale. To offer the Flagship Experience, and reap its benefits, incumbent automakers must become customer-centric.

Customer-centricity is a strategy that aligns a company's development and delivery of its products and services with the current and future needs of a select set of customers to maximize their long-term financial value to the company (Fader, 2012). A *customer journey* describes all the actions and interactions between a customer and a company during their relationship. Some of these interactions are physical, e.g., charging the vehicle; and others are digital, e.g., receiving an offer on the vehicle's infotainment system. Customer-centricity implies finding every way to understand the customer during the customer journey and using each interaction to provide value to the customer because in this way the customer will provide financial value to the company. With so many aspects of our lives being part of our digital existence (e.g., communication, shopping, entertainment, health monitoring, productivity, socialization, and now mobility) customer-centricity requires understanding and being part of each customer's digital life. Understanding each customer means gathering data about the physical and the digital life, and gathering data means instrumenting the customer journey. The more detailed the data captured and the more sophisticated the data analysis, the better the company's opportunities to understand the customer's needs and provide value by addressing them. The *customer experience* determines the value the customer receives and the financial value they provide in return. It encompasses every aspect of the customer journey including the personalization of offerings to address the customer's needs, customer satisfaction with the personalized products and services, and trust toward its brand.

As we have seen in other industries, customer-centric companies enjoy higher revenue growth compared to their competitors with no such practices (Hughes, Chapnick, Block, and Ray, 2021). Customer-centricity leaders such as Starbucks, Ritz-Carlton Hotel Company, and Nordstrom report higher customer loyalty and the ability to capture a larger percentage of the customer's lifetime value than with other revenue-enhancement strategies because they can understand the customer better and rapidly respond to a customer's evolving needs (Huang & O'Toole, 2020). They use every opportunity along the customer journey to showcase their competitive advantage.

The Flagship Experience revolves around the *end-to-end customer journey* that covers the entire relationship between the customer, be it a consumer or a fleet operator, and the OEM starting at the time the customer first interacts with the OEM's brand and ending when the customer no longer owns any of the OEM's vehicles. The relationship may span several vehicles acquired over time. The Flagship Experience also introduces the *cradle-to-grave vehicle journey*. The vehicle journey describes the actions and the interactions between the Software-Defined Vehicle, the OEM, and its partner ecosystem during the vehicle's entire lifecycle, from the time it is manufactured to the time it is retired as it changes owners. For example, a feature that is added to the Software-Defined Vehicle, thus changing its configuration, or a recommendation to remove a feature before reintroducing the vehicle to the market for sale to another customer.

Both the customer journey and the vehicle journey should be owned by the OEM but developed with the collaboration of their partners. A customer journey may be associated with a single customer or a group of customers (or prospects) with shared characteristics. Each automaker can define the end-to-end customer journey with as much detail as it wishes. A journey's touchpoints are established by both the automaker and its partners inside and outside the vehicle. Since the Software-Defined Vehicle is a data generation engine, it is to the advantage of the automaker's partner ecosystem to incorporate in the customer journey as many instrumented touchpoints as possible. The OEM will need to negotiate with its partners about which of the data collected from every touchpoint each party will be able to access. AI plays an important role in both journeys. In the case of the customer journey, it enables the OEM to understand the customer more intimately. In the case of the vehicle journey, it enables the OEM to better exploit the vehicle during its entire lifecycle.

As is shown in Figure 3-1, during the customer journey the customer starts engaging with the OEM at time t0 (the awareness phase of the customer journey) and, if everything proceeds successfully, acquires Vehicle 1 at time t1. The period between time t1 and t3, at which point the customer acquires Vehicle 2, is considered the post-sales period associated with Vehicle 1. During this period, the OEM must try to understand and monetize the mobility-related behavior of every person using the Software-Defined Vehicle, since the customer is not only the person who legally owns the vehicle.

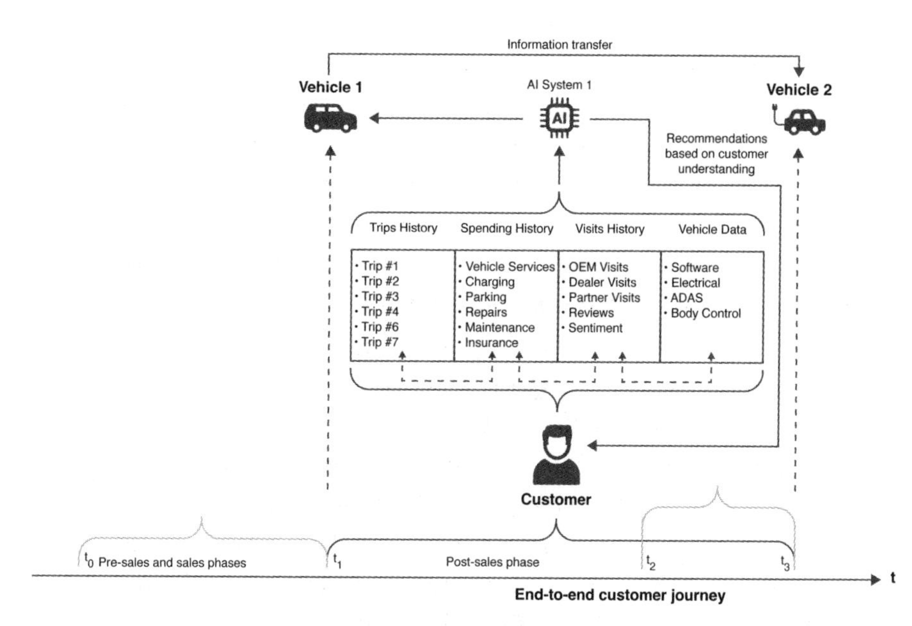

Figure 3-1: *The Flagship Experience's end-to-end customer journey*

Today OEMs pay little attention to the customer's mobility-related activity that occurs during the post-sale phase, particularly once the vehicle is out of warranty, because the customer relationship is owned by the dealer. The OEM starts focusing again on the customer during the next pre-sales effort (t2 to t3) provided that the customer chooses to consider one of the OEM's vehicles. However, to become customer-centric, own the customer relationship, and address the customer's needs for as long as their relationship lasts, they will need to understand the customer's mobility behavior during the post-sales period. For example, what is the typical scenario of a visit to a service center (routine maintenance, repair under warranty, etc.)? What are the various scenarios for vehicle charging (charging at the work location, charging at home, charging while on a long-distance trip, etc.)? For this reason, they will need to capture four types of data.

First, data about every trip the customer takes with the vehicle. When possible, it is also valuable to subsequent analyses to capture the data relating to all the ground trips the customer takes regardless of the modality used because this data reveals customer propensities toward specific modalities. This data enables the OEM to understand how the customer moves and the role the personally owned vehicle plays in this mobility.

Second, data about mobility-related spending. This includes spending on vehicle charging, parking, repairs, insurance, etc. This data not only validates presence in one of the locations captured by the first type of data but also validates the effectiveness of the recommendations offered by the OEM and its partners.

Third, data about vehicle-related visits. For example, visits to the dealer or the OEM's service centers and any reviews associated with these visits. This data shows how loyal the customer stays to the automaker's ecosystem, which itself relates to the opportunities the OEM has to capture the customer's lifetime value.

Finally, the data coming from the vehicle's sensors that relate to the vehicle's status, e.g., battery status and in-cabin behaviors such as the data generated from the interaction with the vehicle's navigation system. This data enables the OEM to monitor the vehicle's state and the cabin occupants' condition and use it to offer recommendations relating to the vehicle's uninterrupted operation, improving customer condition and their enjoyment of the vehicle and reducing the vehicle-related costs. It can also be used by the OEM and its partners to improve service, future vehicle and component designs, etc.

These data types are appropriately linked to create an intricate network that is continuously analyzed using AI system(s) (denoted as AI System 1 in Figure 3-1) to both understand the customer's needs and have an up-to-date picture of the vehicle's performance. The derived understanding is used to tailor the customer experience inside and outside the Software-Defined Vehicle by making mobility-related recommendations to the customer, many of which relate to adding capabilities to the vehicle (e.g., automatic parking) while others, such as extending the Software-Defined Electric Vehicle's range, relate to the vehicle's usage. It is also used to proactively engage the customer during the next Vehicle 1 purchase process, making vehicle recommendations that fit the understanding. When successful, this leads to a shorter sales cycle, but also to increased customer convenience by preconfiguring the customer's new vehicle by transferring from the previous vehicle all the appropriate facilities associated with the customer's digital identity and digital wallet. The goal of these recommendations is for the customer to recognize a benefit whose value translates to loyalty and ongoing monetization.

The cradle-to-grave vehicle journey, shown in Figure 3-2, captures the details associated with each Software-Defined Vehicle from the time the vehicle is first manufactured, including its hardware and software component genealogy, and initially configured by its initial customer, to the time it is no longer available for acquisition by another customer and is removed from the Parc (or fleet).

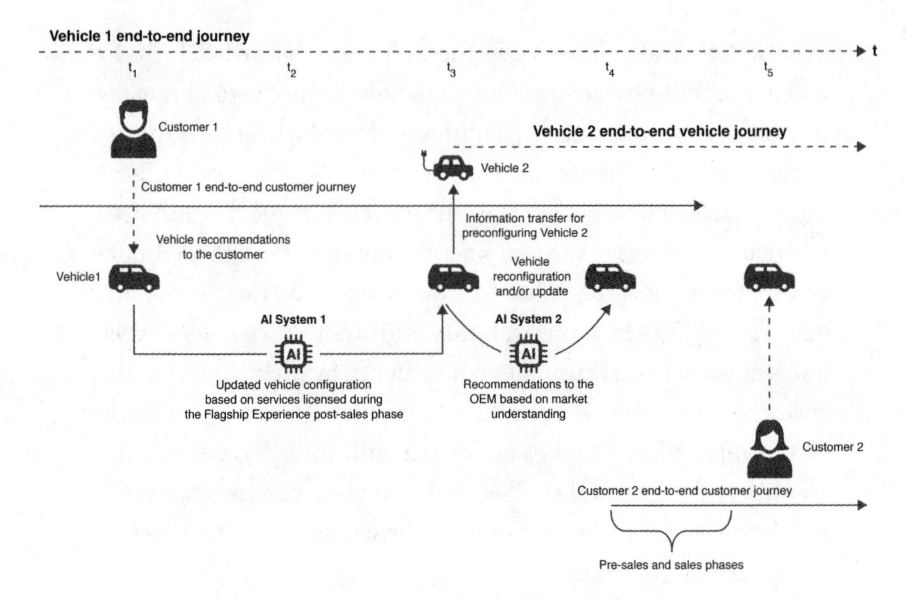

Figure 3-2: *The cradle-to-grave vehicle journey of a Software-Defined Vehicle*

The cradle-to-grave vehicle journey is a unique concept introduced with the Flagship Experience to enable the OEM to monetize the Software-Defined Vehicle over its life and not only when it is first acquired. This monetization is possible because of the software-enabled reconfiguration capability that only Software-Defined Vehicles have. During the vehicle journey, the OEM captures two types of data. First, the vehicle's performance as its configuration is updated due to component updates, repairs, and services licensed or acquired by its current owner/operator. Second, data about the market conditions, e.g., used vehicle inventory, interest rates, used vehicle registrations, etc. When at time t3 the vehicle's current owner/operator returns the vehicle (Vehicle 1) to take delivery of the new vehicle (Vehicle 2), the OEM has the opportunity during the period between t3 and t4 to prepare Vehicle 1 for acquisition by a new customer (Customer 2). The vehicle configuration data collected from each vehicle that is prepared to be reintroduced in the market is combined with the configuration data of other vehicles with similar characteristics, e.g., vehicles of the same model that were operated in the same geography by customers with similar mobility behaviors, and analyzed by AI System(s) (AI System 2 in Figure 3-2) and correlated with the characteristics of the customer segment that will be targeted by the OEM. The AI system utilizes the analysis results to recommend reconfiguring Vehicle 1, by adding or removing features and services, in a way that will increase the probability to be acquired by Customer 2, who belongs to the targeted market segment. In this way, the OEM achieves the ongoing monetization of each Software-Defined Vehicle until the vehicle's end-of-life.

3.1 The Evolution of the Automotive Customer Experience

Until the Great Depression, it was common for the rich to purchase a rolling chassis from an automaker and then go to a *coachbuilder* to develop a car that reflected the buyer's ideas of luxury, performance, or other characteristics they wanted to showcase during driving. The practice started in Europe, where a rolling chassis from automakers like Bugatti, Delahaye, Rolls-Royce, and Hispano Suiza was used as the basis for one-of-a-kind coaches built by firms like Figoni & Falaschi, Mulliner, Castagna, and Spohn. The customer experience was owner-specific and associated with the vehicle's amenities.

Creating bespoke bodies and cockpits on premium chassis evolved into another practice that emerged before the Great Depression in the US: that of the contract builder. Companies like Fisher, Murray, and Budd would build a small number, 100-150, of custom-tailored luxury bodies on order from automakers such as Buick, Hudson, and Packard that would then distribute them through the dealer networks.[3] The small-batch production of luxury vehicles introduced the broader market to the *flagship vehicle* concept.[4] Under this model, the customer experience was also vehicle-centric but reflected the OEM's ideas more than the owner's.

Alfred Sloan, GM's CEO in the 1930s, introduced the *brand ladder* to maintain customer lifetime loyalty as a household moved across life stages and built affluence. The brand ladder led to the establishment of the *flagship brand.* Cadillac is GM's flagship brand. Today each automaker brand develops a flagship vehicle that provides the design cues and language for the brand's entire lineup. The customer experience, and associated customer journey, became specific to the brand but undifferentiated among the members of the populations they targeted. The flagship vehicle figured prominently in print, radio and television advertising, and auto shows to attract prospective buyers to dealer showrooms and keep existing customers loyal to the brand and the manufacturer's family of brands. This evolution of the customer experience remained vehicle-centric.

During the last twenty-five years, the automotive customer journey has become more complex, particularly with the

3 In a sure sign that history repeats itself, when they were first introduced, cars were sold to consumers directly by the automakers. Dealers came into existence about the same time (shortly after 1910) in the US and Europe because of the assembly line. Automakers started producing more cars to meet the growing consumer demand but needed help to sell and support them. Dealers—initially sales agents—were created to address this need.

4 The flagship vehicle should not be confused with the *halo vehicle*. The primary goal of the halo vehicle is to showcase the brand and convince the consumer to buy a vehicle from the automaker's existing lineup, ideally a flagship vehicle. A halo vehicle conveys the automaker's vision, particularly in the areas of design and performance, and even provides hints of features and characteristics that will be found in future production models. It may be a limited-production vehicle like the Chevrolet Corvette C8 Z06, the Mercedes SLR, the Acura NSX, or a concept car like the Mercedes C111 or the Chrysler Atlantic.

introduction of online channels for sales and service and new mobile applications. However, it remains vehicle-centric, is disjointed, and lacks flexibility and adaptability. Its design comes as one of the last processes before releasing the vehicle. The services are not offered in response to a customer's need but because they are available, much like cruise control. Even though they may interact with the prospect during the customer journey's consideration phase, today's OEMs do not have a way to understand each prospect's wants and needs. Even if they did, their vehicles are not upgradable. As Jonah Houston, Ford's ex-director of design strategy, described to the author, "Newcomer automakers are doing a better job in considering the customer experience as they design the vehicle. They take advantage of their Software-Defined Vehicles and design for upgradability by incorporating components that will be used by future services." For example, Rivian incorporated LiDAR into their vehicles even though their software didn't initially support it, as they anticipated developing one or more services that rely on LiDAR and wanted to be able to turn it on using an OTA software update.

To protect various interests, agreements, and norms, and sometimes even address government regulations, there have been few attempts to redesign the automotive customer experience from scratch to connect it with a customer-centric customer journey. To understand the importance of engaging in such a complete redesign, consider an example from the airline industry. Southwest Airlines flies only Boeing 737 aircraft. Over the years it has flown every model of this aircraft. Because of the long-term relationship with the airline, Boeing understands Southwest's needs in ways that other aircraft manufacturers do not—from the price per aircraft Southwest is comfortable paying to the reliability it expects from the aircraft in its fleet, the cockpit configuration to minimize the training pilots must undergo, the type of training services it should offer with every new model, etc. The depth and breadth of this relationship benefit Southwest (its economics, its customer satisfaction score, and other important metrics) and Boeing (helping it improve its aircraft design, manufacturing, and support processes, the instrumentation and sensors with which to equip its planes, and the data it needs to gather from each aircraft).

The result of today's automotive customer experience is lost sales and dissatisfied customers who become less loyal, and whose lifetime value is not fully captured. As Katharina Seifert, head of Volkswagen's Group Engineering Strategy, stated, "Customer experience shouldn't be reduced anymore to the in-car interaction areas but should be thought of as a concept to optimize all customer touchpoints, including the services provided to the customer and the vehicle." To address new mobility's goals and take advantage of the Software-Defined Vehicle's data generation and software-based modification/personalization capabilities, the power of AI, and the extensive role of mobility in customers' digital lives, OEMs must reimagine and redesign the customer experience they offer with these vehicles from the ground up. The Flagship Experience provides the foundation of the redesign.

3.2 Defining the Flagship Experience

The Flagship Experience is a comprehensive customer experience blueprint that customer-centric OEMs can offer with their Software-Defined Vehicles to increase customer loyalty and capture a meaningful percentage of their customer lifetime value by identifying monetization opportunities throughout the customer relationship.

To obtain as clear a customer understanding as possible, the OEM will need to capture mobility-related data both when the customer is using their privately owned vehicle and also other transportation modalities, e.g., public transportation or ride-hailing. Instrumentation will provide the former, while for the latter, the OEM may need to use incentives and partnerships. Recognizing the value of this data, GM is now offering discounts to the owners of its forthcoming Cadillac LYRIQ in exchange for allowing GM to track how they use the vehicle (LaReau, 2022). Tesla is paying drivers to collect data that will be used to improve its Full Self-Driving system (Lambert, 2023). With the instrumented customer journey and vehicle, the OEM captures data from every phase of customer development and during every trip the customer takes with the vehicle. For example, charging the Software-Defined Vehicle during a long-distance trip. In this way, the OEM understands both the customer and the vehicle. The OEM must convince the customer that by providing data, they will be able to receive value, i.e., have a better experience. An initial version of the customer journey may be specific to a customer group with shared characteristics or may be generic to the entire customer population targeted by the automaker. It may only include a small set of scenarios that occur in the course of the customer's relationship with the automaker. With the instrumentation in place through the Flagship Experience, the OEM:

- Collects the data from each instrumented touchpoint, combines it with the mobility data when the customer is using other transportation modalities, and analyzes the resulting data set using AI.[5] The insights generated from these analyses

5 We understand that it may not be practical, feasible, or even desirable to capture every trip that relies on transportation modalities other than the customer's privately owned vehicle. As we describe in Chapter 6, a sample of such trips is typically sufficient. Alternatively, rather than capturing such trips from each individual customer, the OEM can utilize the generalized mobility patterns of a population that exhibits the same mobility characteristics as the target customer.

are used to create personalized value-adding customer recommendations. In Chapter 6 we will describe a customer understanding framework we have developed to generate insights from the customer journey data and the detailed mobility data.

- Presents the generated personalized recommendations, or value propositions. These value propositions range from addressing the customer's particular needs (e.g., providing better safety and convenience through higher levels of driving automation) to adding new services to the vehicle (e.g., improving safety and reducing the Total Cost of Ownership by offering predictive diagnostics). Each of these value propositions takes advantage of the Flagship Experience's and the Software-Defined Vehicle's updatability and configurability. Every proposition should entice the customer to enter into a value-exchanging relationship with the OEM. With some of these propositions, the OEM will monetize the customer. With others, the OEM will gain different types of value. The quality of these propositions impacts the customer's satisfaction, leads to ongoing value exchange between the customer and OEM, and impacts the customer's loyalty toward the OEM.
- Measures and analyzes the results of the customer's action after each recommendation. The customer-centric OEM must always seek feedback about the presented recommendations, even if they are not adopted by the customer, and measure the customer's satisfaction with the adopted recommendations and the overall experience. The analysis of this data using AI leads to a variety of refinements.
- Refines the customer journey (down to the individual customer level), the analyses performed, and the recommendations that can be issued, and repeats the steps. The customer journey is refined by extending it, i.e., by adding new scenarios that

have to be considered, adding new instrumented touchpoints as a result of the added scenarios, and/or changing the instrumentation of existing touchpoints. The extensions must serve the customer better while respecting the customer's preferences and constraints, e.g., the customer's privacy and preferred communication channel. The analyses are refined by adding new data types, changing the way the data is analyzed, or adding to the analysis approaches utilized. The recommendations are refined by changing what can be offered to the customer based on the analysis results, as well as incorporating new services that can be offered to the customer.

This set of steps applies regardless of whether the Flagship Experience is offered to consumers or businesses, though each customer class presents different challenges. In the consumer case, the OEM must address privacy, use of multimodal transportation, multiple data owners, and other issues that can impact the level of customer understanding it can achieve. Competition, including from technology companies that today own the consumer's digital life, can also challenge the effectiveness of the value propositions the OEM generates. The Flagship Experience for Business may initially represent an easier opportunity for the OEM. The data generated by the Software-Defined Vehicle business fleet belongs to the fleet operator, which can provide it to the OEM. Because to date OEMs have not developed a customer experience specific for business fleet operators, and there are few competitors from other industries, the value propositions to be offered under the Flagship Experience for Business will be easier to accept than in the consumer case. However, to develop such value propositions, the OEMs will need to work closely with fleet operators to understand their needs so that they can determine how to address them. As we see from the early success of GM's BrightDrop and Ford's Ford

Pro, today the OEM will have few competitors when offering the Flagship Experience for Business.

Regardless of the customer type, the Flagship Experience must satisfy four requirements. First, it must address every aspect of the mobility-related relationship between the customer and the OEM. This means that the Flagship Experience addresses both pre-sales activities, such as arranging for the time and place for the customer to test the vehicle, and post-sales activities, such as the charging of the vehicle. Since alternative vehicle powertrains are pursued to address climate change, the customer journey should be designed to address relevant customer goals such as acquiring a home charger for the Software-Defined Vehicle and receiving a rebate from the government. According to Jonah Houston, "The industry will be missing a great opportunity if we propagate today's relationship limitations to the new customer experience. The question is whether incumbent OEMs will be able to figure this out."

Second, it must enable omnidirectional, customizable, and value-adding interactions between the customer, the automaker, and its partner ecosystem based on stated and predicted customer needs. These interactions must be predicated on the customer's preferences, including how information should be delivered to them, and on government regulations. Through omnidirectional communication and value-adding interactions, the OEM and its ecosystem continuously update their understanding of the customer's mobility needs and the vehicle's state. Once the vehicle is acquired, the OEM must use the Flagship Experience to continuously provide value by addressing the customer's need for a product, e.g., rear-wheel steering, service, charging station availability or content, or restaurant reviews around a destination. Acquiring new services not only increases the customer's lifetime value but also increases the vehicle's value. The updated understanding can help the OEM and its ecosystem improve processes such as vehicle design, marketing and sales, vehicle manufacturing, and product support.

Third, it must be adaptable and extensible based on the customer understanding gained throughout the customer journey. As the understanding of each customer increases, the customer journey is extended via touchpoints that provide the input for adapting the Flagship Experience. For example, as part of the Flagship Experience for Business offered to a delivery fleet driver whose route has become longer than the vehicle's range, the OEM should recommend that the fleet incorporates a service that predicts when and where the driver should recharge the vehicle so that the deliveries can be completed optimally. Travis Hester, GM's vice president of customer experience, stated to the author, "There are already hundreds of touchpoints that are involved in the customer experience across OEM, dealer network, and their broader ecosystem. These touchpoints were introduced gradually as automotive retailing incorporated digital channels for information gathering, advertising, vehicle configuration, appointments, and many other functions." But as we concluded from this and other interviews with various automaker executives, to date even these touchpoints haven't been integrated under a single customer experience design philosophy.

Fourth, it must be offered with every Software-Defined Vehicle model in the OEM's lineup regardless of class, be consistent across all models, and take full advantage of the vehicle's capabilities. The automotive customer experience offered around today's vehicles varies by vehicle class and model. For example, the customer experience associated with the Lexus UX is different inside and outside the vehicle than the corresponding experience of the Toyota Corolla even though the two are based on the same platform, or between the Lexus ES and Toyota Corolla. While variation in the customer experience may be expected due to vehicle economics, we do not see such differences among the models of the Software-Defined Vehicles from newcomer automakers. The customer experience of Tesla's Model S is the same as that of the Model 3. As we will see in Chapter 4, because of their Software Platform, Software-Defined Vehicles address consistency across vehicle classes without greatly impacting vehicle economics. For another comparison, consider the differences of the customer experience between the cheapest and most expensive iPhone models, or between an iPhone and a MacBook personal computer. While these products have different price points and different features, their pre- and post-sales customer experience is consistent. The inconsistencies in today's automotive customer experience risk customer confusion even when staying within the same vehicle brand. Designing the customer experience from scratch with a common design philosophy will allow the OEM to make it consistent across every vehicle class it offers, even if there are differences in the services offered, e.g., entry-level versus luxury vehicles.

During the ownership of a Software-Defined Vehicle that offers the Flagship Experience, customers will express loyalty by using the products and services provided by the OEM and its partners rather than equivalent competitor products. In this way, the OEM will increase the customer lifetime value it captures. When the customer decides to change vehicles, today the OEM views this as the beginning of a new customer journey because the relationship is between the customer and the dealer. This approach increases the customer acquisition cost and does not capitalize on the knowledge gained during the existing relationship. Instead, as is shown in Figure 3-1, the OEM must view the steps associated with the acquisition of the new vehicle as an extension of the existing journey and capitalize on the understanding gained up to that point to make the right offer, through the customer's preferred channel, about the right new vehicle that will address the customer's stated and predicted needs at the appropriate time. This will increase the probability of having it accepted, increasing the customer lifetime value the OEM captures while reducing the customer acquisition cost. This is the case regardless of whether the customer is a consumer or a business.

This is not the only advantage of the Flagship Experience. It also enables the OEM to optimize the relationship with every owner that takes possession of a Software-Defined Vehicle over the vehicle's life. During the vehicle journey, a Software-Defined Vehicle may have several owners. Each owner will have a distinct customer journey during which he will incorporate his preferences in the vehicle. Consider the following real example. The initial owner of a Tesla vehicle might not subscribe to the Full Self-Driving service, for example, but the vehicle's subsequent owner can still do so. By offering the Flagship Experience with its Software-Defined Vehicles, the automaker must act along two dimensions: the Customer Journey Dimension and the Software-Defined Vehicle Journey Dimension. The customer-centric OEM focuses on the customer relationship but also considers the ongoing connection ("vehicle-centric relationship") with the Software-Defined Vehicle across owners. Of course, for this to be possible, the OEM must own the customer relationship and not the dealer. As was mentioned in Chapter 2, to be part of new mobility's fleet-formation phase, the OEM must also control its Parc. Today the OEM focuses on optimizing the vehicle's monetization as it comes out of the assembly line. In most cases, it does not have a connection with the vehicle once it is out of warranty. The monetization of the vehicle's subsequent owners primarily benefits the dealer. The Software-Defined Vehicle with the Flagship Experience enables the monetization of the vehicle's every owner. As was shown in Figure 3-2, throughout the vehicle journey the OEM can add or remove features and services from a used Software-Defined Vehicle before offering it to a new prospect because it knows the prospect's preferences through previous vehicle ownership and understands the characteristics of a prospective customer segment, or simply to increase or decrease the vehicle's price. In addition, during the pre-sales process of the customer journey, the customer of the used Software-Defined Vehicle may add or remove features both to reflect their preferences and to reduce the vehicle's price.

The Flagship Experience and the corresponding customer journey are co-owned by the OEM's partner ecosystem. The co-owners have a continuously updated understanding of a population's mobility patterns based on how new mobility is adopted, of each customer in this population, and of their mobility needs as they are manifested during the entire customer journey. The co-owners, starting with the OEM, must develop both the necessary technologies and business models. These will clarify the investments, partnership ecosystems, and acquisitions necessary to achieve the Flagship Experience at every point of interaction and facilitate the monetization of the selected propositions. Creating the Flagship Experience involves the OEM, its supply chain partners, its distribution network, and its partner ecosystem. Due to the characteristics of the services and the vehicles that will be part of new mobility, value creation will involve a new set of partners.

An automaker may decide to develop the key technologies required for the Flagship Experience implementation on its own, e.g., Mercedes, in collaboration with partners such as BMW and Google, or license it, e.g., Stellantis from Amazon. The approach the OEM takes may be brand dependent or model dependent. For example, GM can develop Cadillac's Flagship Experience internally, as it did in the past with its Saturn brand, but create Chevrolet's Flagship Experience in partnership with Google. The decision will be driven by financial considerations and the availability of the necessary technology and human factors expertise within the automaker or each of its brands.

The Flagship Experience should position the automaker as the customer's *mobility partner,* not simply as a vehicle provider. Some forward-thinking automakers may transform to become a customer's mobility partner much like Amazon has become the consumer's shopping assistant, and Google has become the consumer's information assistant. Google, Apple, and Amazon are moving beyond their digital-life dominance and are trying to dominate aspects of our physical lives by entering the home and the car. To achieve the mobility partner status, automakers must be able to help the vehicle owner to move *safely, conveniently,* and *affordably* using a variety of modalities, rather than offering just the privately owned vehicle option. The Flagship Experience must present the most innovative mobility capabilities the OEM and its partner ecosystem can offer throughout the customer journey, just like the flagship vehicle offers the automaker's best vehicle technologies. The implementation of the Flagship Experience blueprint will depend on the automaker. Some automakers may elect to implement the blueprint in its entirety, while others may opt to implement only parts of it, e.g., services that can be offered to the cabin's occupants.

3.3 Tesla's Approach to Customer Experience

Tesla has been successful not only because it introduced a new car but also because of its customer-centricity and its industry innovations in vehicle technology, manufacturing, and organizational structure. It brought back vertical integration, which the automotive industry had not seen since the days of Henry Ford, but significantly expanded it by introducing its network of charging stations, and is now considering how to integrate the satellite-based communications offered by Starlink, which is another of Elon Musk's companies. Musk and Tesla also introduced "systems thinking," considering mobility in its entirety with the vehicle as one of its components, and not just the vehicle as a standalone item. Most importantly for this book's reader, Tesla was the first automaker to introduce a well-thought-out novel customer experience around a Software-Defined Vehicle with the associated customer journey. Not only was this an end-to-end customer experience that Tesla could control completely because of its direct-to-consumer model, but, most importantly, the company views it as a dynamic structure that has continued to improve (with occasional issues). Tesla's customer experience starts at the beginning of the consumer journey as the prospect becomes aware of the brand. To date, Tesla does not advertise and does not pay for endorsements. Its global brand awareness has been the result of announcements—including a torrent of well-publicized tweets from charismatic company leader Elon Musk—and mostly positive word of mouth. Lacking the "Elon factor," other newcomers like Rivian and Lucid Motors will not be able to avoid advertising.

Tesla's customer experience continues with the prospect's ability to configure a vehicle online. Key to this capability is the small number of trims, which also enables Tesla to lengthen the effective life of its vehicles. For example, the Model S is offered only in three trims. It is already ten years old with barely a facelift. Following a configuration, the customer can go to a company-owned showroom (in the US states and in the countries where these have been allowed) for testing the vehicle and further refine the configuration. The post-sales experience is enhanced by the in-cabin experience, the company's charging network, the vehicle's servicing, which may include a technician traveling to the customer rather than the other way around, and the functionality of its mobile application, which is well-integrated with the vehicle's capabilities and the overall customer experience. Since the introduction of the Model S, the automaker's mobile application could interact with the vehicle by accessing many of its features, including the Smart Summon feature, which provides the driver with the ability to remotely retrieve the vehicle. It will also have the vehicle autonomously find a parking spot and park itself.

With these capabilities, Tesla can constantly interact with and gather data from each of its vehicles' drivers, establishing a strong bond with them in the process. Tesla fully exploits the data it collects from its fleet of vehicles and uses it to constantly enhance the customer experience and with it the value it offers to the vehicle's user. Data collected from the vehicle is used to assess risk and offer driver insurance at rates that are better than the competition. Location data is key to the ongoing development and prospective deployment of fully automated driving. Tesla's continuously expanding Supercharger network greatly reduced the "range anxiety" associated with battery electric vehicles today for Tesla owners.[6] Informing the driver whether a charger is available in the station and how long it would take to charge the vehicle can help the driver to plan how to spend their time while the vehicle is charging. On several occasions, the company has voluntarily remotely extended the battery range of customers involved in floods and other natural disasters, an act that further increased customer loyalty (Lee, 2018). In this respect, the features of Tesla's vehicles are not additive; they are multiplicative. The customer experience overall has engendered loyalty towards the brand as well as a sense of community among owners, as exhibited in online forums and also around the informal meeting points of charging stations (Libby, 2022).

6 Over the next two years, as more automakers adopt Tesla's North American Charging Standard, their electric vehicles will have access to the same charging network.

Of course, not everything has been smooth sailing. The long wait times to service a vehicle, the lower-than-expected build quality, and the frequently unfulfilled promises (one of the most discussed being around the Full Self-Driving feature) demonstrate the ongoing need to improve Tesla's customer experience.

By capitalizing on the advantages of a deep portfolio of features and services, many of which are installed on the vehicle by the customer after acquiring the vehicle, Tesla has redefined the automotive customer experience into an end-to-end mobility experience. In the process, the company is demonstrating how to establish and benefit from long-lasting and monetizable customer relations both in the US and abroad. It is also demonstrating how to determine which services to develop and own and for which to seek partners. Incumbents like GM, Ford, Toyota, and VW are already starting to embrace and deploy some of these ideas.

3.4 The Flagship Experience in Action

Consider how the Flagship Experience associated with a new Software-Defined Compact Crossover Utility Vehicle (CCUV) can be configured for two different customer segments: the "soccer mom" and the "road warrior."

Assume a young "soccer mom" who lives in an upper-middle-class suburb of a West Coast American city and works in the technology industry. Building upon her existing relationship with an OEM, she starts her research on the OEM's website. Once she makes herself known to the OEM, the OEM must provide her with information relevant to her profile that has been created using the analysis of the data collected from the touchpoints of her customer journey. By employing live agents or AI chatbots, the OEM must use such interaction opportunities to ask questions about how she plans to use the vehicle.

During the Flagship Experience's awareness, consideration, and intent to purchase phases, the automaker:

- Enables her to ask questions and volunteer information and to use every opportunity to verify the inferences that it derived from her mobility patterns. For example, she could ask questions about how frequently to charge and service the electric CCUV. She may also volunteer that she would like to do more "recreational driving" by taking road trips to national parks. The OEM can verify that the customer's commute is short, and she drives alone during "commuting." She "runs errands" (grocery shopping, picking up laundry, etc.) three to four times per week. During these trips, she also "transports kids to sports activities."
- Reassesses her lifetime value to determine offers during the intent to purchase stage. For example, because she will be using the vehicle to transport her children, and potentially their friends, she may be particularly interested in vehicle safety. Two value propositions will be to offer her a data privacy package based on which data generated by the cabin's occupants will *not* be shared with any commercial entity and an insurance rate that reflects the fact she will be using the vehicle primarily for short-distance suburban driving.
- Presents her with appropriate configurations of each vehicle she is considering.
- Instructs the dealer with whom the automaker has connected her to dispatch to her home the vehicle under consideration for her to test.
- Creates a financing plan for each vehicle and includes options for insuring, charging, and servicing the vehicle.

After she acquires the vehicle, the automaker and its partners take opportunities for meaningful interactions with her. Because of her "running errands" and "transporting the kids to sports activities" pattern, the automaker offers an annual subscription to a partner's application that links her children's activities calendar, her to-do list, and traffic information and presents each time the optimal route. In early spring the automaker inquires if she is planning any summer road trips. Upon finding out that she is planning a trip to Zion National Park during her children's spring break, the automaker provides the locations of charging stations along routes to the park and sightseeing side trips, together with a one-time offer to purchase a battery range extension and reconfigure the vehicle's performance to the trip's road conditions. Both capabilities will be delivered via an over-the-air update of the vehicle's software. The automaker uses the data being collected from the vehicle to infer that it is being driven by two individuals with different driving styles. Based on each driving style, the automaker makes real-time recommendations on where to charge the vehicle as the trip is unfolding. In this respect, the automaker thus becomes the customer's mobility partner. The vehicle becomes another enabler of the customer's digital life, as the smartphone and tablet are already. As GM's Hester stated, "The vehicle and the smartphone become one through connectivity." The battery range extension increases the vehicle value. In this respect, the vehicle's value is constantly increasing rather than decreasing, which is the case with the current vehicle acquisition model.

A different Flagship Experience will be associated with the "road warrior" who does not already have a relationship with the OEM and is considering a pre-owned Software-Defined CCUV. Based on questions asked by the OEM's chatbot, we find that the prospect has an office in the middle of a European city, but her home is in a distant suburb. She "commutes" to her office once a week using her privately owned vehicle because of the unavailability of convenient public transportation options. She does not have parking privileges at her work location. During the other workdays, she uses her vehicle to travel to office locations around the broader metropolitan area. Shopping is delivered to her home, and she never goes on road trips. When she takes her vehicle for service, she expects a quick turnaround time but also the availability of a loaner vehicle.

After assessing her lifetime value, the OEM proposes pre-owned Software-Defined Vehicles whose range can be extended using a software update and includes a top-of-the-line infotainment system. The value propositions include a discounted parking-and-charging package in a facility nearby her office, a monthly subscription to an anti-theft protection application that utilizes the vehicle's external cameras, an offer to have the vehicle picked up from her home whenever service is needed along with the simultaneous delivery of a loaner vehicle, and recommendations on the changes she will need to make to her home's electrical system to charge the vehicle together with recommendations to partner local home solar system installers and charging station networks. Due to her busy schedule and lifestyle, the OEM brings the vehicle to her home so that she can test-drive it. She opts for a lease and completes the entire transaction online.

After taking delivery of her vehicle and observing her driving style, the automaker presents her with an offer that includes auto insurance at a discount over her current carrier, an annual subscription to the highest level of driving automation system offered by the automaker, and an offer for the advanced powertrain package that changes the vehicle's steering and braking responses, making them sportier. The vehicle enhancements will be delivered via an over-the-air update of the vehicle's software. Upon accepting the offers, her account is charged automatically, and she can choose to either pay for the driving automation and powertrain enhancements as a one-time transaction or, together with the insurance offer, as monthly subscriptions that can be added to her lease payment. In this way, the automaker and its partners establish a value-exchange relationship with her while showcasing the brand's differentiation by anticipating the customer's needs using the observed behavior and addressing them effectively. The customer receives value through better insurance coverage and her vehicle's upgraded capabilities. If these capabilities are appreciated, the customer's reciprocal value is in the monetization of the offer and the loyalty toward the OEM's brand.

Through mobility, customers want providers to address convenience, safety, pollution avoidance, and economics, including the vehicle's Total Cost of Ownership (TCO), and superior overall user experience. In the examples provided above, the soccer mom cares about safety while transporting her children but also about conveniently accomplishing her errands. The road warrior is looking for convenience and better economics whenever she visits her office, but also by having her groceries delivered to her home. She wants to install a solar charging system at her home because she cares about climate change. As the customer realizes that the automaker addresses their mobility-related needs by recommending capabilities that increase the value and enjoyment they derive from the vehicle, their relationship with the automaker transforms.

3.5 Capabilities, Features, and Services

The Flagship Experience includes capabilities, features, and services provided by the OEM and its partners to address a customer's needs. Some of these are about the customer, such as the capability to inform the prospective customer about a model's possible configurations. Others are about the vehicle, such as extending the

Software-Defined Vehicle's range or configuring the powertrain's sensitivity (comfort at one extreme, racetrack at the other).

A set of these capabilities, features, and services may be available across all Software-Defined Vehicles an OEM offers because they are what we call **brand-defining**. Large screens became a Tesla brand-defining feature. The goal of these is to establish the customer's connection with the automaker's brand. Additional features and services may be specific to a model. We call these features and services **model-defining**. Heated seats are a BMW iX model-defining feature. The goal of these is to establish the customer's connection with the vehicle. Brand-defining and model-defining features and services are typically bundled in the Manufacturer's Suggested Retail Price (MSRP).

Features and services that provide value beyond this bundle warrant a fee. We call these **personal features or services**. The usage-based insurance or the anti-theft protection an OEM may offer to the customer are personal services, and a new gauge look-and-feel, as Ford recently offered in the 2024 Mustang Mach-E, is a personal feature (Schrader, 2022). Some of the model-defining and personal features and services are selected at the time the vehicle is configured during the sales process because the hardware they require may already be installed by the automaker. For example, the vehicle may already be equipped with a self-park capability at sale time. Other personal features are selected after the vehicle is acquired in the course of being used. Some of the model-defining features and services and many, but not all of the personal ones, are monetizable. For example, Ford's gauge look-and-feel was acquired after the vehicle was sold and was paid using a one-time transaction. Others, such as GM's OnStar or Tesla's insurance services, are monetized using subscription models. Several other business models are possible.

When interviewed by the author, executives from incumbent automakers acknowledged that they need to find new ways to monetize the customer after the vehicle is acquired. As Ola Källenius, Mercedes-Benz CEO, stated "Do I think we can make money with it [the vehicle]? Yes. Do I know exactly how much? No, but the potential is there." Incumbent OEMs are starting to experiment with post-sales monetization, but no results are being reported yet. The financial markets prefer to see corporations use subscription business models because of their recurring revenue characteristics. For consumers, the preference for one-time transactions versus monthly subscriptions depends on several different factors, including the amount to be paid, the perceived long-term utility of the product or service, and others. Alternative business models will lead to different consideration criteria.

OEMs must address two questions in designing each personal feature or service. First, how much effort will be expected from the customer? Second, what will motivate the customer to personalize their experience by adopting it? Consumer electronic devices, such as smartphones, have many personalization options, but most people don't take advantage of them either because they don't understand them or because they don't derive any value from them. In some cases, consumers may spend an hour setting up their device but never make any changes afterward. The key to answering these questions is to establish a continuous value-exchanging relationship with the customer.

Features and services are associated with:

1. **Safety and personal security**. For example, a pedestrian detection service uses data from cameras and LiDAR to automatically detect pedestrians unaware of the vehicle's presence and can stop the vehicle without human intervention, thus avoiding an accident.

2. **Privacy and cybersecurity.** For example, a data management service that determines what data to transmit from the vehicle to the OEM's databases by using the privacy-related policies and permissions the vehicle's owner has defined for each member of the household and issues alerts every time an outside entity tries to access data stored in the vehicle's systems, thus protecting the owner's privacy.
3. **Economics (time savings or money savings).** For example, a navigation service that provides delivery drivers with detailed navigation instructions of where to safely park the vehicle, which building entrance to use, and how to move within the building where the delivery is made, thus enabling a delivery driver to make more deliveries per work period and avoid receiving parking tickets.
4. **Convenience.** For example, a service providing access through the infotainment system to an office productivity suite, such as the VIIZR suite developed by Ford in partnership with Salesforce, that enables the vehicle's user to securely record or access information from the company's back office systems, thus increasing productivity (Weintraub, 2022).
5. **Entertainment.** For example, an entertainment service like the one provided by Netflix uses the vehicle's large video screens, haptics, in-cabin sensors and actuators, and superior audio system to deliver a cinematic experience within the cabin.

A survey of 3,000 individuals conducted in Germany, the US, and China by PwC's Strategy& group revealed that customers rank safety, convenience, and vehicle management (relating to money savings) highest among classes of connected services (PwC, 2020).

I foresee four types of monetizable transactions within the Flagship Experience. First, transactions directly between the customer and the OEM. Second, transactions between the customer and the dealer. In this case, the OEM receives a percentage of each such transaction. Third, transactions between the customer and a partner from the automaker's ecosystem. Again, the OEM receives a percentage of this transaction, albeit smaller than the percentage received from the second transaction type. Fourth, transactions between the OEM and a partner benefit the customer but the customer does not pay. These may be one-time transactions or subscriptions. And, of course, the monetization of the consumer is quite different from the monetization of the fleet owner/operator. Each requires distinct strategies.

OEMs may use their direct customer relationship to offer monetizable capabilities, features, and services through a marketplace they operate directly or license from a supplier. Tesla takes the former approach, whereas Mercedes, Polestar, GM, and Renault the latter. Mercedes, BMW, VW, Stellantis, and other OEMs use the marketplace that Faurecia developed through a joint venture with Aptoide (Guillaume, 2022). Polestar, GM, and Renault license different configurations of the marketplace that is part of Google Automotive Services. Alternatively, the OEM may establish a dealer network to sell vehicle-, driver- and passenger-centric applications and services. Such a network would resemble the networks that mobile telephone providers like Verizon and AT&T have established in addition to the retail stores they operate.

Because they are not customer-centric and consequently lack the deep customer understanding associated with customer-centricity, today most incumbent OEMs offer features and services that are not easily monetizable. Some charge for brand-defining and/or model-defining features or services that customers expect for free. For-fee services introduced by automakers like Toyota, BMW, and VW to date have achieved significantly lower adoption and revenue compared to the OEM projections. This may be the result of offering low-value features and services and/or the result of a poorly executed pricing strategy. Toyota tried to transition its connected vehicle remote start feature to an eighty dollar annual subscription (Stumppf, 2021). Based on comments posted on online forums and feedback from the company, few of the eligible connected vehicle owners signed up for the subscription. Many owners pushed back because they did not consider this supposedly time-saving service valuable enough to be worth a subscription. Its user experience was poor, taking thirty to sixty seconds for the remote lock/unlock to activate. Some consumers did not see enough value to pay for this service at any price. They considered the remote start a brand-defining service that must, therefore, be offered for free. A pricing strategy that offered three or four years of connectivity-related features upon vehicle purchase might have been more acceptable to these owners. In a survey of 4,640 individuals conducted by IHS Markit in 2021, 14 percent of the responders indicated that they would not renew their subscriptions to connected vehicle services, with 5 percent of the responders stating that they didn't find the services valuable (Davis, 2022). These results demonstrate again that most automakers lack the requisite customer understanding to create high-quality services.

Three big questions for the industry are: where will the monetizable opportunities come from, who will create the services to capture them, and who will ultimately benefit from them? Will it be the OEM, the suppliers, the dealers, a third party like Google, a startup, or a different type of organization? GM's Travis Hester believes they will come from the automakers. Ford's Jonah Houston believes they will come from technology companies. Google, Apple, and Amazon provide in-cabin technologies, starting with the infotainment system but also extending beyond it (Shah, 2022; Nellis and White, 2022; Huntin, 2022; and Mims, 2022). These companies have already demonstrated their ability to introduce and monetize their customers and are aggressively extending their digital life solutions to the vehicle. Nikos Michalakis, senior engineering fellow at Woven by Toyota, believes they will come from new application development studios (Michalakis, 2021). OEMs will compete with these technology companies in customer experience adoption.

To succeed with the Flagship Experience, incumbent automakers will need to innovate at every level from organizational and business models to technological transformations. The most important technological ingredients: the Software-Defined Vehicle, its software, and AI applications. Because of their architecture, Software-Defined Vehicles support the frictionless collaboration among all the parties involved in providing the Flagship Experience and the cloud infrastructure that manages the generated data to facilitate the vehicles' monitoring, management, and updating. The Flagship Experience requires many software and AI innovations that are detailed in Chapters 5 and 6, but the Software-Defined Vehicle, described in the next chapter, is the key technology for making it a reality.

CHAPTER 4

The Software-Defined Vehicle

On June 1, 2022, Ford's CEO declared that "software and electrical (i.e., digital) architectures (are) the heart of the vehicle transition" that is driving the automotive industry's radical transformation. The Software-Defined Vehicle provides major benefits to the automaker and is the right vehicle platform for offering the Flagship Experience to the customer. Today all existing and announced Software-Defined Vehicles are electric. Over time, OEMs may decide to use Software-Defined Vehicle architectures in conjunction with other powertrain types, e.g., hydrogen. But for our discussion, we assume only Software-Defined Electric Vehicles. In this chapter, we introduce Software-Defined Vehicles, provide a brief overview of vehicle architectures, and show how these vehicles enable the Flagship Experience.

A Software-Defined Vehicle is a vehicle whose functions are primarily enabled through software. Software-Defined Vehicles differ from conventional vehicles over five distinct dimensions. They enable continuous improvements of the vehicle and its in-cabin experience through over-the-air (OTA) software updates, support higher driving automation, enhance the electric powertrain through intelligent battery management, enable the control, optimization, and management of a fleet, and facilitate the vehicle's interaction with the transportation infrastructure. A prime example of a Software-Defined Vehicle capability is the self-parking function of Tesla vehicles.

A survey conducted by IHS Markit revealed that 70 percent of consumers who intend to buy a new vehicle expect it to have OTA update capabilities that will be used to enhance its features, increase its safety, and provide added conveniences (IHS Markit, 2020). Tesla addressed problems associated with the braking, window defroster, and pedestrian warning signal components of its Software-Defined Vehicles simply through OTA software and firmware updates without requiring the vehicle owners to visit a service center. In addressing these issues expeditiously, Tesla improved its vehicles' safety, provided convenience to its owners, and impacted their productivity. The ease with which these problems were resolved was not simply a function of Tesla's OTA software and application update capability. The architecture of Tesla's software and hardware platforms made the remote updating of the software easy. The architectures of Software-Defined Vehicles will be reviewed in the next section.

Today, every time that any part of a conventional vehicle's software needs to be updated, particularly concerning a recall, the dealers must be involved. This is an expensive and time-consuming process for both the automaker and the customer. The OTA updating that is supported by the Software-Defined Vehicle forgoes the need for many of these visits. A study conducted by BearingPoint in Germany showed that 43 percent of vehicle recalls could be addressed using OTA updates (Penthin and Landgrebe, 2019). OEMs also expect to deliver via OTA updates on most of the revenue-generating services they will offer to customers. However, they still need to develop business models to ensure that this will be net new revenue rather than replacement revenue to make up for lower revenue from vehicle sales.

Due to their architectural characteristics and the use of electric motors rather than internal combustion engines, Software-Defined Vehicles over time will become easier and cheaper to manufacture than conventional ICE vehicles. On average an electric vehicle's powertrain consists of 60 percent fewer parts than the powertrain of a vehicle equipped with an internal combustion engine. Having fewer parts means that the vehicle is easier to manufacture with fewer points of potential failure during the assembly process. Manufacturing advances, combined with fewer parts, lead to lower per-unit manufacturing costs (Lambert, "Tesla releases video of Giga Press in action producing giant single-piece rear body," 2021). Tesla is already demonstrating that the simplicity of the Software-Defined Vehicle results in lower manufacturing costs (Lambert, "Tesla achieved supply optimization at Gigafactory Shanghai, now says cost is linked to materials," 2020). Software-Defined Vehicles are expected to reach manufacturing cost parity with conventional vehicles before the end of the decade (Miller, 2020; Partridge, 2021). BCG estimates that by the end of the decade, automakers will see a $600 to $2,500 cost improvement in the Software-Defined Vehicle's BOM, depending on the vehicle class (Koster, Arora, and Quinn, 2021). As Ford's CEO recently stated, build-to-order, a process greatly facilitated by Software-Defined Vehicles, will be adopted globally and will help the company better control manufacturing costs and improve margins (Foote, 2021). Taking out the battery cost, which is decreasing due to new battery chemistries, and because of the high software content and use of commoditized electronics components, Software-Defined Vehicles have a lower hardware-related Bill of Materials (BOM) cost. However, they will have a more expensive software-related BOM because the type of software used in these vehicles, including the AI software, is more expensive to develop and test. Generative AI systems such as GitHub's Copilot and OpenAI's GPT-4 can be used to reduce the code-development cost, which increases the productivity of software engineers, but we are still in the early stages of evaluating the potential impact of these solutions (Warren, 2023; Capelle, 2023; Eloundou, Manning, Mishkin, and Rock, 2023).

Fewer hardware parts combined with increasing manufacturing automation leads to higher-quality vehicles. Higher-quality vehicles lead to lower warranty accruals. Higher-quality vehicles imply that the automaker's warranty costs, and consequently the necessary accruals, can be lower. In 2016, Tesla was accruing 3.8 percent of its revenue for warranty costs, while that amount was reduced to 2.6 percent by 2021. Overall, during the same period, GM was accruing 2.2 to 2.4 percent, with the percentage increasing to 5.8 because of the battery problems of the Bolt EV, while Ford was accruing 1.6 percent to 3.6 percent during the same five-year period. In 2019 automakers spent 2.9 percent of their revenue on warranties. On average, for conventional internal combustion engine vehicles, automakers are accruing $539 per vehicle sold.

While once production is set up the cost to manufacture Software-Defined Vehicles will be lower than the equivalent cost of ICE vehicle, the transition from conventional vehicle architectures to Software-Defined Vehicles will be expensive for both the incumbent OEMs and their suppliers. This is due to the technological, organizational, and business model transformations that will be necessary during this transition. These transformations and their implications are discussed in Chapter 8. The transition will be completed in phases, and at times will be difficult. As we are already seeing from the announcements by automakers such as GM, VW, Ford, and others, the investments that will be necessary to develop and release such vehicles will be extremely high

(Lienert and Bellon, "Global Carmakers Now Target $515 Billion for EVs, Batteries," 2021).

4.1 Microcontroller-Based Architectures

The Software-Defined Vehicle is not about the introduction of software into the vehicle. Today all vehicles incorporate software. Some models include more than 150 million lines of software, often more software than advanced aircraft. However, in today's vehicles, the software is mostly in the form of low-level microcode and used as "glue" for the vehicle's hardware. Microcode (or firmware) is embedded among more than one hundred hardware-based Electronic Control Units (ECUs) developed by suppliers. ECUs are organized into complex and closed architectures, which we call *Microcontroller-Based Architectures.* They communicate mostly via low-bandwidth local networks, called Controller Area Networks (CANs), and occasionally via Automotive Ethernet and Low Voltage Differential Signaling in the case of the infotainment system (Wikipedia).

Microcontroller-Based Architectures have been adopted by automakers since the seventies for two reasons. First, they enable OEMs to focus on powertrains while Tier 1 suppliers develop the rest of the technologies required by the vehicles in the model lineup. For example, the vehicle's body control or its infotainment system. Over time, Tier 1 suppliers like Bosch, ZF, Denso, and Magna built separate divisions to specialize in each of these areas and provide the necessary components to OEMs like Mercedes, Stellantis, Toyota, and BMW. Second, they enable the functional safety processes to certify systems and subsystems with fixed functionality to be considered as black boxes independent of one another. As a result, the overall vehicle is certified via a dependency diagram.

Adding new functionality to a vehicle based on a Microcontroller-Based Architecture is a slow and error-prone process because it involves updating the microcontroller's firmware. This can only be done using specialized equipment operated by knowledgeable technicians. As its name implies, the primary function of a microcontroller is to control. While these components have computing capabilities that over the years have increased, computing is used to *accomplish the control function.* Microcontroller-Based Architectures have evolved, been patched, and been augmented incrementally from vehicle generation to generation and even model year to model year. Many of the functions that we associate with advanced technology vehicles, such as the Advanced Driver Assistance System (ADAS), are surpassing the capabilities of these architectures and require the Software-Defined Vehicle.

4.2 Software-Defined Vehicle Architectures

The Software-Defined Vehicle is characterized by the type of software it uses, the clear separation between hardware and software, and the role its software plays in the vehicle's upgradability and customizability. As Scott Miller, GM's Vice President of Software-Defined Vehicles, stated to the author in March 2022: "The Software-Defined Vehicle provides the necessary foundation for the OEM to offer a customizable user experience because of the architecture it is based on." The Software-Defined Vehicle relies on an architecture that separates the software from the hardware and creates functionality using applications, similar to how personal computers rely on operating systems to abstract away their hardware from user-facing applications. Several of the mechanical components found in conventional vehicles, such as steering or braking, can now be replaced by electronic components that are controlled by application-level software rather than just firmware.

The architecture's principles are based on modern High-Performance Computing (HPC). Under this architecture's purest form, which we call Central Computer Architecture, the vehicle's functions, from simple ones, such as opening and closing the cabin's windows, to complex ones, such as managing the electric vehicle's driving automation or its battery management system, are software applications that are running on the vehicle's operating system and middleware. Like every other computer system, the vehicle now gets a full-function operating system. The middleware and the operating system comprise the vehicle's *Software Platform.* The details of the Software Platform are provided in Chapter 5. All software is running on a high-performance central computing server that interacts with the vehicle's storage, sensors, and actuators via messages that are communicated using a broadband Ethernet network. The computing server consists of multiple CPUs, GPUs, and specialized AI processors that can perform neural network operations. All these components comprise the vehicle's *Hardware Platform.* The vehicle's batteries, motors, and associated components that make the vehicle's movement possible are integrated into the vehicle's *skateboard,* which acts as the electric vehicle's chassis. The skateboard encloses the vehicle's batteries, its electric motors, and other components. A high-level view of this architecture is shown in Figure 4-1.

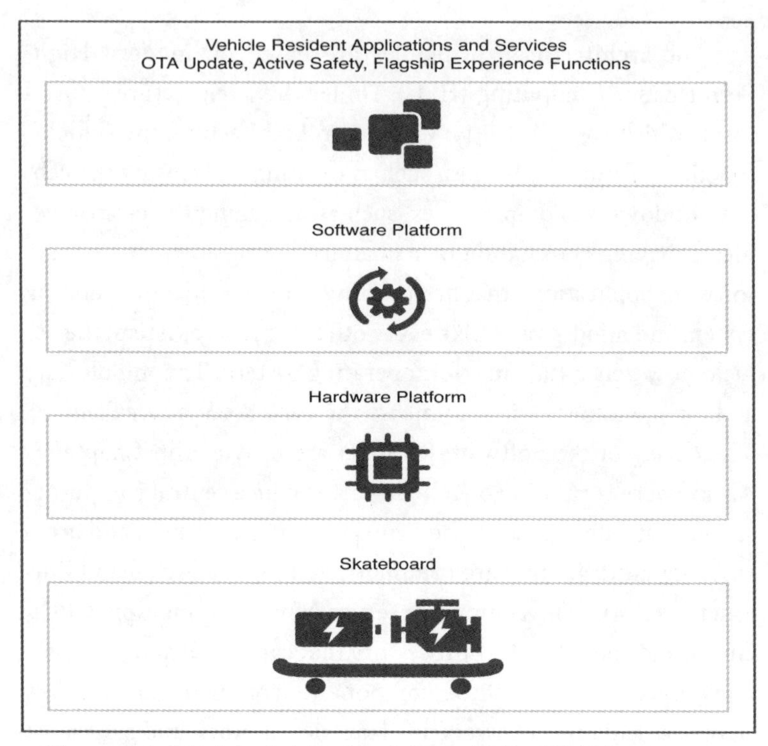

Figure 4-1: *A high-level view of the Software-Defined Vehicle architecture*

The skateboard and the hardware platform comprise the vehicle's E/E Architecture. The data that is generated by the architecture's components is accessed through well-defined APIs. This data describes the state of the vehicle, the state of the cabin, and its occupants. When coupled with the appropriate AI applications, some of which may be running on the vehicle while others are in the OEM Cloud or even the cloud infrastructures of the automaker's partners, it can be used to predict the vehicle's needs based on its current and expected operating environments. Chapter 6 describes how the captured data is exploited with the use of AI, enabling the OEM and its partners to offer a different Flagship Experience to each customer.

The Software-Defined Vehicle architecture:

- Makes possible the vehicle's continuous improvement using OTA updates, enhancing its performance, and personalizing the customer experience.
- Supports higher levels of driving automation, enhancing safety and convenience.
- Makes the writing, testing, and incorporation of the new software, and the potential upgrading of the hardware, easier, faster, and cheaper because of standardization and component commoditization and reuse.
- Manages the performance of the electric powertrain.
- Manages and optimizes the performance of a fleet (a corporate fleet or the Parc), which is important for the customer experience.
- Enables the vehicle's easy and cost-effective interaction with the transportation infrastructure, also enhancing safety and overall experience.

Incumbent OEMs will evolve their Software-Defined Vehicle architectures over several versions during this decade and next, much like the architectures of today's Internal Combustion Engine vehicles have done over several decades. Each version will optimize around specific criteria such as time to market, battery material availability and cost, technology development costs, software and hardware component costs, vehicle-operating characteristics such as power consumption, the ability to attain specific manufacturing goals, and the cost to service the vehicles sold. The Microcontroller-Based Architectures used by incumbent automakers today will be followed by Domain-Based Architectures and then by Zone-Based, and finally Central Computer Architectures, as shown in Figure 4-2.

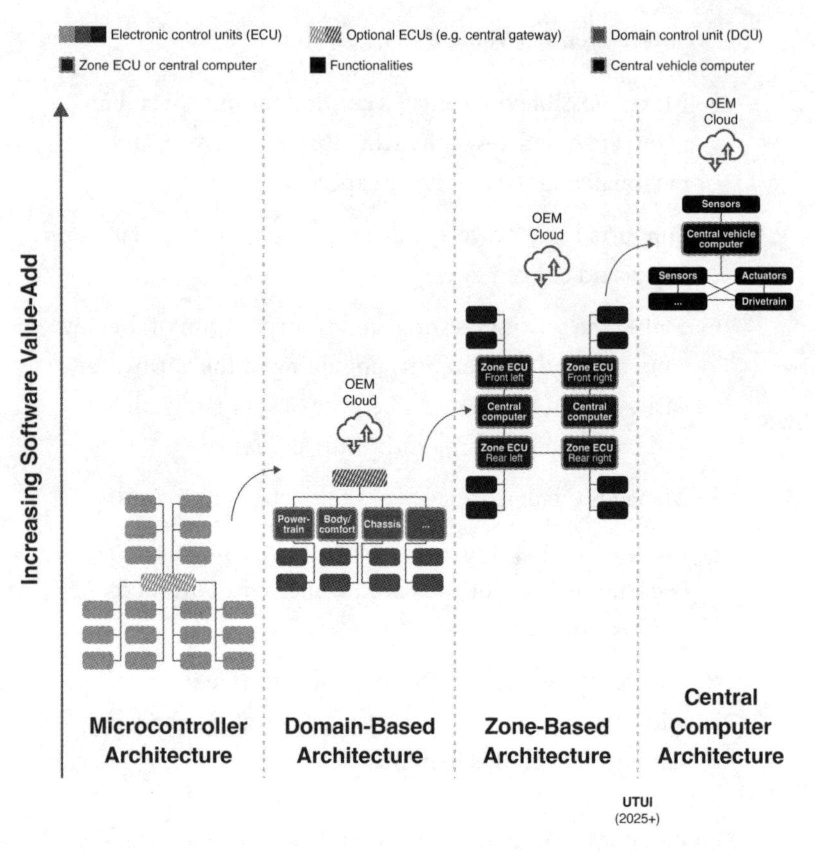

Figure 4-2: *From Microcontroller-Based to Central Computer Architectures*

In a Domain-Based Architecture, each group of related ECUs is organized into a separate domain controlled by a Domain Controller (these are hierarchical structures). Domain-Based Architectures include at least four to five domains: powertrain, ADAS, infotainment, body control, and passive safety. A typical Domain Controller has a CPU that performs the necessary domain functions and provides a simplified interface to other Domain Controllers. Domain Controllers communicate via a high-speed network. The ECUs controlled by the Domain Controller perform lower-level functions.

The hardware platforms of VW's ID.4 and Ford's Mach-E use a Domain-Based Architecture. Domain-Based Architectures involve fewer ECUs than microcontroller architectures. For example, the Ford Mach-E has 51, and the VW ID.4 has 52, whereas a typical vehicle that is based on a microcontroller architecture uses at least one hundred ECUs. The domain controllers are updated over the air, whereas the ECUs they control continue to require visits to service centers.

Domain-Based Architectures represent an evolution from microcontroller architectures, but a less than radical one. They are still difficult to manage because of their complexity and often create conflicting sets of requirements since every domain defines requirements within its scope. Software-Defined Vehicles available today from incumbent automakers use Domain-Based Architectures because their implementation doesn't require major disruptions to the existing supply chains and manufacturing processes. The Domain Controllers are typically provided by the incumbent Tier 1 suppliers used by the OEM, as we are already seeing in the case of VW, Mercedes, and Toyota. Among others, a typical Domain-Based Architecture includes a domain controller for active safety, e.g., ADAS, a controller for the vehicle's infotainment functions and in-cabin experience, and a third for controlling the vehicle's body, e.g., the vehicle's suspension. Domain controllers also don't have a major financial impact on the vehicle's hardware-related BOM. By splitting the vehicle's computing load among lower-priced Domain Controllers, the OEM can offer lower-priced electric vehicles. As a result, it is possible to use a Domain-Based Architecture to offer basic ADAS functionality (lane-keeping, traffic sign recognition, blind-spot detection, and automatic braking) in its mid- and lower-priced vehicles.

Interviews with incumbent OEMs and Tier 1 suppliers working with these architectures revealed that their organizational structures remain largely the same as they transition from microcontroller to Domain-Based Architectures. All-in-all, they are the least-disruptive path toward the short-term implementation of Software-Defined Vehicles. But they are not the choice for OEMs in the longer term.

In a Zone-Based Architecture, a computing server, called the zone controller, controls a set of components (ECUs, sensors, and actuators) based on their physical location in the vehicle. For example, the sensors, actuators, and ECUs handling ADAS-related functions together with the corresponding components relating to the body control functions for the vehicle's entire front area are handled by one computer server. Zone-Based Architectures require fewer but more powerful controllers (zone controllers) to process the data and perform the complex operations associated with functions such as driving automation, battery management, and others. These controllers are developed specifically for these architectures. The hardware platforms of the Software-Defined Vehicles from Tesla and NIO use Zone-Based Architectures. For example, the Tesla Model Y has 26 ECUs, and the NIO ES8 has 35 (Kane, 2021; Jost, 2021). Each of the computers used in Tesla's architecture has a peak performance of 73 TOPS (WikiChip, 2019). The computer servers of Pony.ai's autonomous vehicles have a peak performance of 254 TOPS (Tao, 2022). NVIDIA, whose CPUs and GPUs are part of many such servers, has announced that its next-generation processor targeting autonomous vehicles will provide 1,000 TOPS of peak performance (Wiggers, 2021). As is shown in Figure 4-2, incumbent OEMs are not expected to introduce vehicles using Zone-Based Architectures before the 2024 model year. For example, Mercedes will introduce a new software platform, MB.OS, in 2024 for vehicles that will be based on a

Zone-Based Architecture, whereas Ford's GE2 platform is expected to be introduced around 2024 and to be paired with the Zone-Based version of FNV in 2025.

Zone-Based Architectures have several advantages that make them appropriate for fulfilling the promise of the Software-Defined Vehicles and the Flagship Experience. Because they use fewer controllers, they require simpler wiring architectures. A simpler wiring architecture leads to lighter vehicles. Weight has a direct impact on an electric vehicle's range. The components of these architectures are designed from the ground up to be updatable over the air, resulting in a process that is simpler and faster to perform and manage. Even hardware upgrades or replacements are easier to perform. As we are already seeing from the Software-Defined Vehicles introduced by newcomer automakers, it is possible to easily replace a sensor or sensor suite or to upgrade a controller's capabilities with additional memory and new CPUs and GPUs. More of the functionality is implemented in software that can be updated. This means that overall the vehicle can be updated and reconfigured more easily and at the lowest level than vehicles that are based on Domain-Based Architectures throughout the vehicle journey. Because of their hardware and software architecture, these vehicles generate more easily accessible data. This data is not only valuable for achieving the Flagship Experience's promise, but it can be used in the Software-Defined Vehicle Journey to improve the vehicle's ongoing exploitation by the OEM. Because of their power, the computers used in Zone-Based Architectures can process more data in the vehicle, e.g., performing data filtering and compression, applying machine learning, and encrypting data. As a result, less of the data generated by the vehicle's sensors needs to be transmitted outside the vehicle. This capability not only reduces the communication costs between the fleet and the automaker's cloud but also reduces the automaker's cost of cloud-based data storage and the risk of data leaks during data transmission while protecting occupant privacy. Finally, because of their higher application software content, these architectures enable easy customization of the vehicle to the needs of a market. For example, the brand-specific, model-specific, and personal services offered as part of the Flagship Experience could be different in vehicles sold in the European Union market versus the same models sold in the US market. Recently, Tesla filed a patent application for software-controlled headlamps that can conform to any global regulatory standard (Fox 2022).

As the computing power increases while power consumption and costs decrease, Zone-Based Architectures will be followed by Central Computer Architectures. In these architectures, the computers managing Zone Controllers are replaced by a single central computing server. As is shown in Figure 4-2, by the end of this decade, clean sheet Software-Defined Vehicle architectures will have evolved into the Central Computer Architecture. In Central Computer Architectures even more of the vehicle's functions will be handled through software, leading to additional savings for the automaker and a better, continuously improving Flagship Experience for the customer. The vehicles that will be based on this architecture will also support higher levels of driving automation—i.e., safer vehicles—and include a larger set of applications to manage the vehicle's more powerful sensors.

Regardless of the architecture they are based on, Software-Defined Vehicles are data factories. They must therefore incorporate advanced safety, cybersecurity, and data privacy strategies, which require:

1. Designing safety, cybersecurity, and data privacy at the systems level rather than the ECU level, as with current generation vehicles.

2. Changing the functional safety model because the vehicle's logic is moved to the software platform and is decoupled from the hardware it affects.
3. A certification process that spans the entire vehicle lifecycle, from the start of production to the OTA updates of deployed vehicles. As we are starting to see with Tesla, governments will also need to become involved in the development of acceptable OTA update processes that can be at least standardized globally (Bellon, Jin, and Shepardson, 2022).
4. Adopting both an approach for the proactive testing of the *entire vehicle,* as well as a reactive approach whenever a threat is recognized after the vehicle is acquired. This will require new testing and simulation technologies, detailed in Chapter 5, to identify and assess the impact of potential threats. Today the testing of vehicles that are using Microcontroller-Based Architectures is a slow process that costs hundreds of millions of dollars, requiring many test vehicles to be developed and retrofitted until the start of production.
5. Simultaneously monitoring the vehicle's software and hardware components for cybersecurity threats that can impact both the vehicle's safety and its occupants' privacy.
6. Taking advantage of the OTA updating capability to immediately respond with the appropriate remediation approaches.

4.3 Software-Defined Vehicle Development

A Software-Defined Vehicle may be developed using a "clean sheet approach" or a "retrofit approach." In the retrofit approach, a Microcontroller-Based Architecture is transformed into a Domain-Based Architecture. The vehicle's hardware platform is paired with a simpler Software Platform to enable the OTA updates, certain ADAS features such as automatic braking and lane-keeping, and maybe a low level, e.g., L2, of driving automation. The Ford Mach-E crossover SUV was developed using the retrofit approach. Its architecture is based on a variation of Ford's GE1 hardware platform and its FNV software platform. GE1 is a derivative of Ford's global compact (C2) vehicle architecture.

The clean sheet approach starts with Zone-Based Architectures. The vehicle's Software Platform and E/E architectures are built from the ground up. Vehicles such as Tesla's Model S, GM's LYRIQ, NIO's ES6, and XPeng's G9 are developed using the clean sheet approach. While every Software-Defined Vehicle, regardless of the development approach used, can utilize the Flagship Experience, vehicles based on the clean sheet approach a) generate data of finer granularity and greater variety that can be accessed using more sophisticated APIs, and b) are more configurable and can take better advantage of the Flagship Experience's services through software delivered to the vehicle using OTA updates or to channels preferred by the customer.

The adopted approach depends on how quickly the OEM wants to introduce to the market vehicles with features requiring a Software-Defined Vehicle architecture, how it plans to evolve the architecture, the cost of adopting a particular architecture, the organizational readiness of the OEM and its supply chain, and the capabilities—e.g., OTA updates and driving automation—the OEM wants the Software-Defined Vehicle to have. In Mach-E's case, Ford designed an electric vehicle with OTA update capability that can be used on certain software components, e.g., the dashboard. The retrofit approach enabled Ford to start producing the Mach-E within a year of announcing it. On the other hand, in addition to producing an electric vehicle that would be capable of receiving OTA updates, GM wanted the Cadillac LYRIQ midsize

SUV to use a new battery technology and be capable of Level 3 driving automation. For this reason, GM decided to use the clean sheet Ultium/Ultifi Software-Defined Vehicle platforms. However, due to this decision, GM has only been able to start LYRIQ's production in 2023.

With the clean sheet approach, OEMs develop simultaneously the vehicle's software and hardware platforms. Scott Miller indicated that GM, together with its partners, is simultaneously developing Ultifi, the Software Platform, and Ultium, the E/E architecture. VW is following a similar approach. During our communication, Dr. Jutta Schiffers, SVP responsible for the vehicle operating system at VW's CARIAD, stated that VW is simultaneously developing One Digital Platform, the problem-plagued software platform, and MEB, the hardware platform of its Software-Defined Vehicles (George, 2022).

There are several advantages of the clean sheet over the retrofit approach, which include:

1. Easier evolution of the vehicle's architecture as computing power increases.
2. Shorter follow-on product development times with more robust hardware and software platforms because these are developed under a single design and implementation philosophy shared by all participating partners.
3. Opportunity to introduce more applications in and around the vehicle on the same platform because of the cleaner separation between software and hardware.
4. API-based control at the hardware architecture's lowest level, resulting in easier OTA updates for the entire vehicle.
5. Lower manufacturing cost when production scale is reached, particularly when combined with gigapresses and/or additive

manufacturing,[7] increased manufacturing speed, and the opportunity to upgrade hardware as technologies improve (Johnson J., 2022).

6. Better long-term economics for the OEM, its supply chain partners, and the vehicle's operator because of more frequent, easier, and cheaper-to-deploy updates, resulting in higher margins for the OEM and better ROI for the operator.
7. Better in-vehicle data management because of the improved ability to control and filter the data in the vehicle before it is transmitted to the cloud and improved utility and economics of the cloud-based systems.
8. A better implementation of the Flagship Experience.

Incumbent automakers are often ambivalent regarding how to proceed with clean sheet Software-Defined Vehicles. Clean sheet approaches require larger investments for the development of Zone-Based and eventually Central Computer Architectures, hiring people with the necessary skillsets in areas in which OEMs have not traditionally excelled, such as modern platform software, semiconductors, materials, and battery chemistries, organizational and culture changes, and the radical rethinking of the manufacturing process and the partner ecosystems. The technologies incorporated in Zone-Based Architectures have never been produced at scale by incumbent OEMs or their Tier 1 suppliers. While incumbent OEMs will continue to heavily rely on their long-term Tier 1 suppliers (note the ongoing work between Toyota and Denso, VW with Continental and Bosch, and Mercedes and Bosch around

7 Tesla focuses on manufacturing operations capitalizing on its Software-Defined Vehicles' architecture as much as it focuses on its customer experience. For example, Tesla brought its per-vehicle cost down from $84,000 in 2018 down to $36,000 in 2022 by improving its vehicle designs and factory designs. GM is now trying to do the same.

Software-Defined Vehicles), new partners will provide the necessary software, semiconductor, and cloud computing technologies. Companies such as NVIDIA, Qualcomm, AWS, and KPIT are quickly emerging as prominent members of such networks. GM has announced a partnership with Qualcomm, VW with Microsoft and NVIDIA, and Ford with AWS (Nellis, "General Motors taps three Qualcomm chips to power its Ultra Cruise feature," 2022; Novet, 2019; and Wheatley, 2019). However, while demand for electric vehicles is increasing, it is not yet clear that customers are willing to pay the higher price Software-Defined Vehicles currently command, which makes OEMs ambivalent about how much to invest and how much to transform. We discuss these issues in Chapter 7.

4.4 The Software-Defined Vehicle's Impact on the Flagship Experience

We opened this chapter by stating that the Software-Defined Vehicle is the vehicle platform around which to offer the Flagship Experience. The sophistication of the Software Platform, particularly in the Zone-Based and Central Computer Architecture variants, provides a level of configurability to the vehicle and also a variety and quantity of data that enables automakers to personalize a customer journey down to the individual or customer segment, instead of creating a generic experience for all models of a brand. A good example of this opportunity is provided in the Honda/Sony Software-Defined Vehicle that is marketed under the brand Afeela (George, "Sony and Honda's EV goes where the Apple Car never did," 2023). Its in-cabin experience is optimized around multimedia consumption and is designed to engage the customer with movies, music, and video games.

During the pre-sales phase the Software-Defined Vehicle's Software Platform, particularly that used in a Zone-Based or Central Computer Architecture, combined with the AI systems of the Customer Management Platform and the Vehicle Management Platform that will be described in Chapter 6, enables the prospective customer to create the vehicle they need and want (vehicle features and services), rather than having to buy one that is in the dealer's lot and whose features have been selected to address the perceived needs of broad market segments. The prospective customer can easily test different vehicle settings and features before ordering the model they want to acquire, transforming the traditional test-drive into a more comprehensive evaluation experience.

Because these vehicles are equipped with more sensors and software components than conventional vehicles, they will also generate data of finer granularity, which is made easily accessible via their Software Platform, combined with the Hardware Platform. This data can be analyzed and exploited throughout the vehicle's journey, from the time it is designed to the time it is retired, even as it goes through multiple owners and operators. It benefits everyone that participates in the value chain: the OEM, suppliers, sales and distribution channels, service providers, and certainly the vehicle's owners and operators. More and better-quality data are key ingredients to AI machine learning systems for monitoring, diagnosis, and performance optimization. As we will see in Chapters 5 and 6, some of these systems are onboard the vehicle as part of the Software Platform and others are in the OEM Cloud. They are used to improve the vehicle's manufacturing processes and performance, provide better customer service and support, and also optimize the vehicle engineering and manufacturing processes. Using the data and the output of specific AI systems, the automaker and the members of its partner ecosystem can monitor how the customer interacts with the vehicle and make suggestions on features the customer may appreciate, or simply enjoy. The vehicle is dynamically configured, down to the vehicle-, driver-, and passenger-centric applications and services, while being tested, and the customer orders exactly what they want and feel they need. The resulting personalization of the vehicle leads to ongoing customer monetization.

The data generated by these vehicles can be used to reduce the cost of ownership by optimizing the vehicle-related costs to its observed use. Understanding the driving style, its impact on the driver's risk profile, and the places where the driver parks the vehicle are equally valuable to the private vehicle owner and to the fleet operator assessing drivers and the impact on insurance rates and financing options and rates. By being in the position to directly assess the driver's risk profile due to their driving style, the automaker can offer competitive insurance and financing rates, further increasing the customer lifetime value. The results of these analyses can also be used by the automaker to determine when to suggest a vehicle change or attending a driving school, maximize the customer's ROI, or recommend vehicle maintenance to minimize maintenance costs.

The data that is generated about every aspect of the vehicle, down to the smallest software and hardware component, combined with the OTA software updating capability and OTA frictionless payments, can enable many types of vehicle repairs without having to visit a service center. In this way, the customer is happy because their vehicle remains safe and roadworthy, and they do not experience any productivity loss or inconvenience. The automaker is satisfied because they have controlled the vehicle's repair (positively impacting its residual value) and directly monetized the overall experience. Optimizations of the vehicle's performance can result in extending the range of an electric vehicle without increasing its battery capacity, extending the life of the tires, or even minimizing steps required to complete a set of tasks, e.g., deliveries, which is particularly important for the operators of logistics fleets. All these are monetizable opportunities that are part of the Flagship Experience. In addition to Tesla, GM, Ford, and Mercedes have announced their intention to offer post-sales services for their Software-Defined Vehicles to increase customer lifetime value and loyalty.

The vehicle's configurability due to its Software and Hardware Platforms, together with the generated data and the AI systems that use it, enables increasing the return on the customer's investment through hardware and software upgrades over time. The return on investment may be in the form of improved productivity, convenience, safety, increased residual value, and even entertainment and environmental impact. As was detailed in Chapter 3 and is further explained in Chapter 6, the highly instrumented Software-Defined Vehicle enables the OEM and its partners to continuously monitor and analyze the way the vehicle is used by the fleet's drivers or the members of the customer's household. Equipped with such analyses, the automaker and its partners can proactively recommend new capabilities to be added to the vehicle. In this way, the customer derives increasingly higher value from the vehicle, and the automaker and its partners derive higher value from the customer. The Flagship Experience can be tailored to the individual customer even if they are not the vehicle's first, or even second, owner. Furthermore, by using the data generated by the customer and the vehicle, the automaker can identify new features and services that will need to be developed for their vehicles, determine how to engage with its partner ecosystem, recruit new partners, and remove existing ones as appropriate. Over time, the Flagship Experience that is enabled by the Software-Defined Vehicle enables the automaker to build a highly differentiated value chain.

The Software-Defined Vehicle's extensibility and upgradability combined with the AI-based analyses of the generated data enables the OEM to monetize the vehicle over its life, the Software-Defined Vehicle Journey mentioned in Chapter 3, by offering a different but equally monetizable Flagship Experience to each of the vehicle's owners. Extensibility and upgradability result in an increase in the vehicle's residual value. As was mentioned in Chapter 3, the acquirer of a Microcontroller-Based used vehicle must live with the choices of the vehicle's first owner, which are the dealer's choices. This ability to reconfigure used vehicles can lead to slower product obsolescence, as well as enable the automaker's financing arm to offer lease terms with lower monthly payments. Moreover, because they will continuously collect data from their Software-Defined Vehicles, OEMs will be able to participate more easily in used vehicle sales, including sales among private parties.

How the automaker decides to approach the Software-Defined Vehicle directly impacts the type of Flagship Experience it can provide. While today we see most OEMs developing proprietary Software-Defined Vehicle architectures, it is not clear how many of their efforts will survive. Closely tying the Flagship Experience with the Software-Defined Vehicle architecture will create stronger reasons for survival. The evolution of smartphone architectures and the connection between the strongest ones and the customer experiences they provide could serve as an important example for the entire automotive industry.

CHAPTER 5

Software for the Software-Defined Vehicle

Software is the key ingredient of the Software-Defined Vehicle and what makes possible the Flagship Experience. As the vehicle architectures progress from Domain-Based to Central Computer, the software content in and around the vehicle increases, as does its complexity and the types of AI used to manage the vehicle and the services comprising the Flagship Experience. Incumbent automakers may be experts in large-scale manufacturing, together with their suppliers in the embedded software used in Microcontroller-Based Architectures. But they have a lot to learn about the development and deployment of the software platforms and applications that will define the Software-Defined Vehicles and the Flagship Experience offered around them. This chapter presents the software used in the operation of the Software-Defined Vehicle. The software used to implement the Flagship Experience taking advantage of the vehicle's software is discussed in Chapter 6.

Two recent announcements demonstrate the importance of software in Software-Defined Vehicles but also the difficulty incumbent automakers have in developing and deploying this software. Volvo announced that it will delay the introduction of its EX90 software-defined SUV because it has not completed the software system that is used to test the vehicle (Rasmussen, 2023). The delay is also impacting Volvo's Polestar brand, as its SUV uses the same architecture as the Volvo SUV. Volkswagen announced that the next version of its software platform, being developed by its CARIAD organization, will be delayed by at least two years and won't be ready until 2028 (Waldersee, 2023). VW's ongoing software problems led to the dismissal of its Group CEO in 2022 but also to the dismissal of CARIAD's CEO (George, 2022). This software makes possible the operation of the Software-Defined Vehicle. So, the three-year-old CARIAD is now on its third CEO. Software problems are not plaguing only incumbents but newcomers as well (Butters, 2023).

These examples show that no matter how well-designed the Flagship Experience is, if the Software-Defined Vehicle does not have the appropriate software underpinnings, it will not be deployed. The Software-Defined Vehicle relies on three types of software: the Software Platform that makes possible its operation; applications, such as ADAS, that enhance the vehicle's basic functionality; and the Software Tooling Platform which includes the tools used by all the processes before the vehicle being delivered to the market. Certain applications that are part of the Flagship Experience run on the vehicle, providing the in-cabin experience, while others run on the driver's smart device. These applications rely on the capabilities of the Software-Defined Vehicle's Software Platform, as well as on software running on the OEM's cloud (see Figure 5-1). To create these systems, automakers and the members of their ecosystems must adopt a new software development process. In this chapter, we discuss the development process and present the Software Platform and the parts of the Software Tooling Platform that relate to the Flagship Experience.

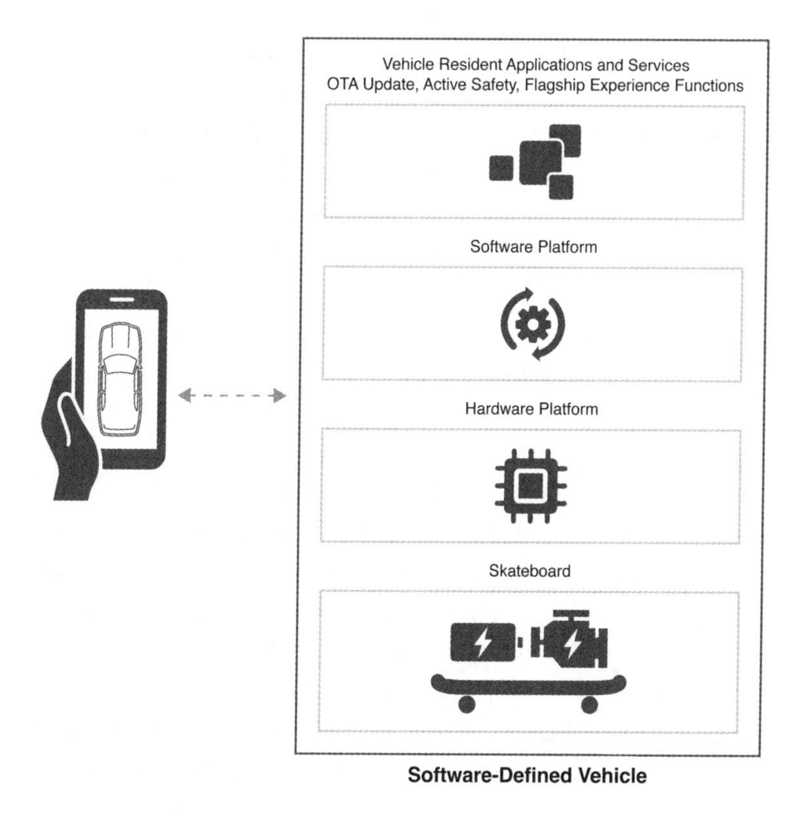

Figure 5-1: *Software relating to the Flagship Experience is running in the Software-Defined Vehicle, the OEM Cloud, and the customers' mobile devices.*

5.1 The Software Development Process

The Software-Defined Vehicle and the Flagship Experience are tightly linked, such that all the software needs to be built as a comprehensive and well-integrated system that can be leveraged across all vehicle models and incrementally improved over time. This will be a new practice for automakers, but one followed routinely by high-tech companies. Traditionally, when developing a new vehicle, automakers start bottom-up by first designing and building a vehicle. They then fit the software, most likely developed by one of their suppliers, around it, typically covering the in-cabin interactions. The software is different from model to model, resulting in high costs for the automaker. It is also clunky and disconnected, resulting in experiences that dissatisfy and frustrate customers.

To achieve its software goals, the automaker must establish consistency across all software (a "software philosophy"), develop a detailed architecture for the entire software system, and adopt an agile software development process for its implementation. As part of this process, it must enforce DevOps practices and Continuous Integration/Continuous Delivery (CI/CD) tactics. The DevOps practice consists of eight steps: plan, code, build, test, release, deploy, operate, and monitor. This practice must be applied not only during the vehicle's initial development and release but throughout the entire vehicle journey. CI/CD relies on *software tooling*, which will be discussed later in the chapter.

The need for a unifying consistency across the Software-Defined Vehicle and the Flagship Experience means that the overall architecture governing the vehicle and experience must:

- Simultaneously encompass all the vehicle's layers (skateboard, hardware platform, and software platform) and the services that will comprise the Flagship Experience, and
- Unify data, software, hardware, APIs, and security.

The software architecture must not only enable configurability, extensibility, and upgradability but also inform the automaker's other processes (manufacturing, support, servicing, etc.) that benefit from understanding the impact of every modification of the vehicle's initial configuration because of fixing a problem or adopting, removing, or upgrading a service.

5.2 The Software Platform

The Software Platform is what makes the vehicle Software-Defined. As Doug Field, Ford's chief advanced product development and technology officer, stated during the company's May 2023 Capital Markets Day, "It is the software platform that will differentiate the vehicle and not its hardware platform." In Chapter 4 we provided a high-level overview of the vehicle's architecture. Figure 5-2 shows the Software Platform's major components in more detail. This platform consists of middleware, a real-time operating system, the infotainment operating system, and APIs.

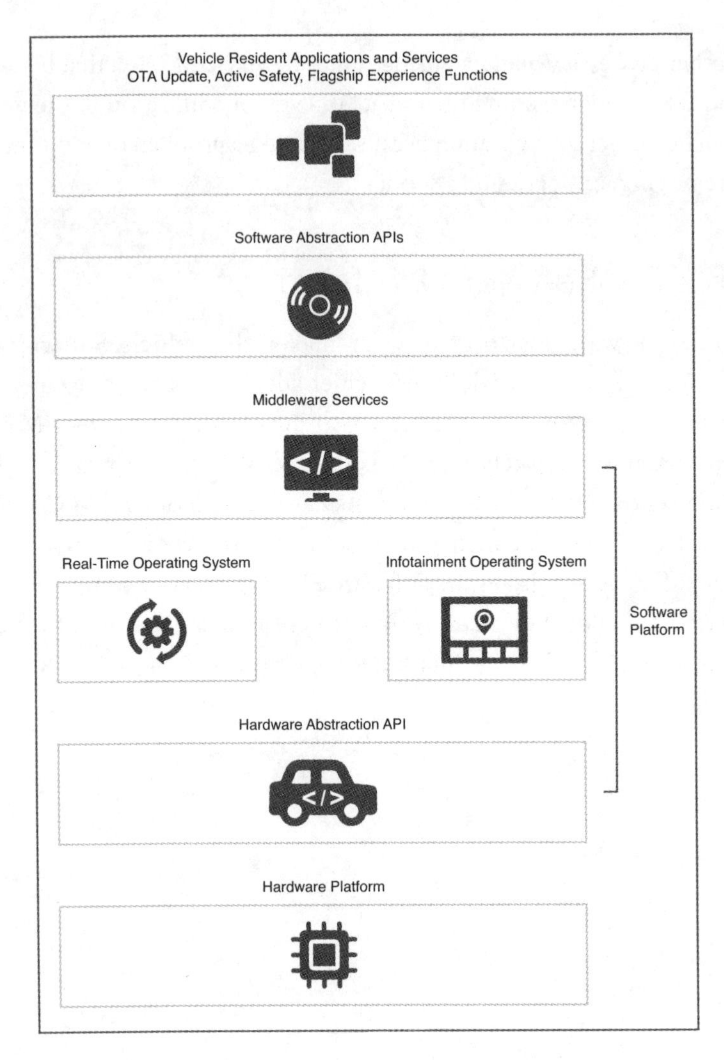

Figure 5-2: *The components of the Software Platform*

The approach automakers take with the Software-Defined Vehicle's operating system depends on hardware cost (operating systems run on a CPU and require memory), the cost to develop or license the operating system itself, the complexity of the hardware managed by the operating system (e.g., a zone controller is a complex hardware piece and more complex than a domain controller), and the level of the required functional safety that must be provided by the operating system. In Domain-Based Architectures, each controller runs on a separate instance of the real-time operating system. Many OEMs developing Software-Defined Domain-Based Vehicles are adopting the QNX real-time operating system to manage the domain controllers because it has received functional safety certification and runs on a variety of CPUs. GM has adopted the Red Hat Linuxsystem, which has also received such certification. For the infotainment operating system of their Software-Defined Vehicles, some OEMs are also using QNX, while others are adopting Google's Android Automotive.

Because of virtualization and the higher computing power of zone controllers, in Zone-Based and Central Computer Architectures, each controller will be able to run multiple instances of the operating system. Furthermore, if containerization is adopted by these architectures, we can foresee the development of multiple specialized operating systems, e.g., an operating system optimized for ADAS and another optimized for the electric powertrain, that can run on the same controller. We can also foresee the possibility that Tesla may agree to license its real-time operating system to other OEMs that develop Zone-Based Architectures in the same way it recently licensed its charging adapter design.

In general-purpose computer systems, middleware is used to provide the services that applications need, but the operating system does not include it. Operating systems such as QNX provide only kernels, i.e., they support the key functions needed for the software to access the hardware. For this reason, it is necessary to develop middleware to support the operating systems and the applications running on the vehicle. The functionality of the Software Platform's middleware layer can vary. In some cases, e.g., ZF's middleware, it may be used to provide communication and connectivity among the applications that access the Software Platform. In other implementations like GM's Ultifi, Mercedes's MB.OS, Hyundai's CC.OS, and Toyota's Arene, the middleware has broader functionality connecting the vehicle to the OEM's cloud, managing the OTA updates, and connecting the vehicle with the OEM's partner ecosystem. It also includes AI applications that perform tasks, from optimizing the battery's performance to determining which data to send from the vehicle to the OEM Cloud, as well as invoking the right cybersecurity function to protect the vehicle and its occupants. To be able to successfully update and upgrade the vehicle's capabilities, as well as introduce new capabilities to it, the middleware must include a complete software-based model that includes the vehicle's hardware components and their connectivity. This implies that every partner whose components are incorporated into the vehicle's architecture must provide the corresponding model of each component. Such a complete vehicle model does not exist today. Moreover, it is not clear whether each partner will agree to provide such a model to the OEM.

For the architecture to simultaneously encompass all three of the vehicle's layers, it must rely on APIs that provide access to the functions and signals determined by the automaker. As automakers move from Domain-Based to Zone-Based Architectures, this access becomes more fine-grained to ultimately encompass every one of the vehicle's sensors and actuators. APIs enable the OEM and its partners to continuously upgrade the vehicle's components, e.g., upgrading a Domain Controller or the application responsible for automatic braking without concern they will stop performing as intended, provided that the API of the changed component remains the same. A Software-Defined Vehicle architecture combined with the right API architecture will ensure that most of the updates and upgrades can be performed over-the-air, but even the ones that require hardware, and thus visits to the dealer, such as the upgrade of a sensor, a CPU, or GPU, can be performed quickly with minimal labor involvement.

The Software-Defined Vehicle could incorporate several applications and services that make possible its operation and provide it with capabilities that differentiate its performance or comply with regional or global regulations. Some of these applications will be developed by the OEM and others by its partner ecosystem. The Flagship Experience applications running on the Software-Defined Vehicle depend on the Software Platform's functionality, i.e., on the services provided by the adopted operating systems, the services implemented by the applications provided as part of the Software Platform, and the richness of the APIs.

The Software Platform of vehicles that use Zone-Based Architectures will be more complex than the platforms of vehicles that use Domain-Based Architectures. This is because Zone-Based Architectures have fewer ECUs, more sensors and actuators, and more applications. The zone controller's software, starting with the operating system, must now perform the functions performed by the software of the removed ECUs. The applications will require more sophisticated middleware and a richer API architecture. The Software Platform enables virtualization, i.e., it creates multiple virtual environments (VMs) on a single physical hardware platform. This allows its software components to operate in their isolated virtual machines. Virtualization enables controller consolidation, improves security, and improves the Platform's performance.

The Software Platform can be implemented in a variety of ways. A preferred way is by using a service-oriented architecture (SOA), in which the desired functionality is provided by microservices. A services-oriented architecture facilitates distributed software development and updating, as well as portability across vehicle models. Implementing the Software Platform using a services-oriented architecture will require that the Software-Defined Vehicle is equipped with high-performance CPUs and high-bandwidth Ethernet-based networks. However, the Software Platform can also be implemented in a more centralized manner to improve system performance, particularly when the hardware platform includes less powerful CPUs.

A microservice is a small independent service. An application can consist of a collection of microservices that are accessed using APIs. Microservices communicate over the architecture's Ethernet-based network. A microservice may be developed and owned by an individual or a team. Microservices have several advantages. They are scalable because each microservice runs independently, making it easier to add, remove, or update it. The result is systems with higher fault tolerance. Each microservice can be developed using a different programming language. This enables the selection of the most appropriate language for the function the service needs to perform and takes advantage of the development team's competencies. Microservices are simple to deploy, allowing the addition of functionality without disturbing the rest of the architecture. They provide strong data security because a system's components are broken into smaller pieces that can be connected via secure APIs, ensuring that data is only accessible by authorized users and applications, as well as being compliant with specific standards such as GDPR.

5.3 The Software Tooling Platform

Software-driven companies like Google, Amazon, Meta, Apple, and a few others have developed software tools for designing, prototyping, debugging, and deploying software and machine learning models, as well as for managing the agile development process. These software tools have been integrated into *tooling platforms.*

Combined with the adoption of agile product development processes, these platforms are considered major contributors to the higher engineering performance of these companies over their competitors, and to the innovations they bring to market. By integrating software and data, these platforms enable distributed teams to work effectively, pay attention to details that the members of a team may otherwise omit, ensure that corporate quality standards are maintained, manage APIs properly, enable software to be deployed consistently to internal and external customers, and measure everything from a team's productivity and work-quality to a customer's use of each released product. Customer feedback is quickly reflected in new product releases, leading to customer satisfaction and loyalty. The continuous and consistent use of these platforms by all product teams leads to results that corporations in other industries envy. It results in products that customers make part of their digital lives and help refine with their feedback. Hardware tooling has been part of the automaker's vehicle design and development process. For example, the Toyota Production System that resulted in the automotive industry's second radical transformation required the development of tooling.

Realizing the benefit of this approach and because they view themselves as technology companies rather than pure automakers, newcomers like Tesla and Rivian adopted similar practices and developed impressive software tooling platforms (Singh, 2021). The incumbent automotive industry, along with several other industries, is either in the very early phases of adopting similar processes and tooling platforms or has not yet started. The Software Tooling Platform is part of the OEM Cloud. Though some of these tools may also be used for the development of the Flagship Experience, e.g., the Digital Twin described below, the vehicle requires a rich set of tools whose description is outside the book's scope.

The Software Tooling Platform must address five requirements. First, it must support the entire vehicle journey, from initial design and validation to the vehicle's last OEM-led update configuration and refurbishment. The Software-Defined Vehicle is governed by three lifecycles:

1. **The software lifecycle,** with OTA updates of applications and services every four to six months and a major release every year. This is the release cadence of most enterprise software vendors. Newcomer OEMs are already achieving this cadence today.
2. **The sensor/consumer electronics and battery technology lifecycle,** with hardware updates every eighteen to twenty-four months. Over the past several years, the consumer electronics and IT industries have developed this practice.
3. **The skateboard and body lifecycle,** with three-year updates for the vehicle's body and interior, and seven-year—or even longer—updates for the skateboard. Consider that the Tesla Model S body is over ten years old whereas the Model 3 body is over six years old. Both vehicles continue to sell well globally. Even if the battery technology is updated, the vehicle's skateboard may not need to be updated as frequently.

During a Software-Defined Vehicle's journey, the OEM updates the vehicle's software, and as it moves from owner to owner it may choose to completely reconfigure the software and the applications already installed in the vehicle by its previous owner. The goal of such a reconfiguration will be to reflect the OEM's updated market understanding and the anticipated needs of the customer segment to which it will market the vehicle. Depending on how the vehicle is designed, during the vehicle journey the OEM may also reconfigure and update the vehicle's sensors, batteries, and certain electronic components. The updated skateboard and body can only be installed as the vehicle is being manufactured.

Second, the tooling must include software for managing the agile development process. These tools must enable flexibility while maintaining organizational structure. They must track each project while organizing the team's progress, making it easy for developers to achieve their goals. This will help each team identify requirements and break them into tasks. Given that an important part of this process is communication, these tools must support discussion and planning.

Third, it must handle the complexity of programming, testing, and validating the entire Software-Defined Vehicle. DevOps combined with the microservices-based implementation enable program testing and validation for the entire Software-Defined Vehicle

Fourth, pipelines between the Software-Defined Vehicle and the OEM Cloud must enable the quick flow of data and software. This flow will ensure that the product teams can assess the vehicle's performance.

Fifth, as was mentioned above, the architecture must unify software, hardware data, APIs, and security. The software tooling infrastructure must support the integration of legacy applications and systems since at least some of the tools already being used today by the automaker will carry forward to the Software-Defined Vehicle.

The development of the Software-Defined Vehicle requires, among others, the following tooling: a data platform, digital twins, intelligent simulators, a machine learning stack, a software and machine learning model version control system, and Integrated Development Environments (IDEs) for vehicle applications and services. The tooling may run on the OEM's private cloud or on hybrid clouds to facilitate the collaboration of the automaker's distributed teams, as well as the teams that are part of the automaker's worldwide partner ecosystem. In this section, we focus on tools notably relevant to the Flagship Experience.

5.3.1 Digital Twins

A digital twin is an accurate representation of a physical object, complete with the processes that govern the object's operation. Typically, the representation of a digital twin is continuously updated as the physical object is enhanced and the understanding of its components and processes improves and is refined. A digital twin may exist for small and simple objects, like a pen, or extremely complex systems, such as a campus or even a city. A digital twin is used in conjunction with intelligent simulators (described in the next section) and other AI systems to understand the behavior and assess the performance of the physical system it represents. Oftentimes, the digital twin is built before the physical system is developed. It is then used with the associated simulator to understand system shortcomings, design defects, and unforeseen situations that will need to be addressed before the physical object is materialized.

For the implementation and deployment of the Flagship Experience, it is important to develop a digital twin of each Software-Defined Vehicle in the OEM's lineup. The twin will utilize the Vehicle Model, i.e., a model of the complete vehicle. The Vehicle Model must be represented in several different levels of detail because each level is used for a different purpose. Versions of the Vehicle Model are used by the Software Platform's middleware and the AI system described in the next chapter. This implies that it will be advantageous for the OEM to have a digital twin for each Software-Defined Vehicle in its Parc. Through the model, the OEM will be able to assess the impact of each action that can be performed in the course of offering the Flagship Experience. The OEM may decide to build the digital twin on its own or incorporate models provided by the suppliers of the cabin's components, e.g., seating. In addition to developing a digital twin of the cabin, it may also be advisable to develop a digital twin of physical parts of the Customer Journey (e.g., a digital twin of the charging of the vehicle) to simulate the potential interactions between the customer and a touchpoint (e.g., the charger. The clean sheet Software-Defined Vehicle is an ideal use case for digital twins. The vehicle's architecture, its high software content, and the digital components that comprise even its hardware systems facilitate the digital twin representation.

5.3.2 Intelligent Simulator

The configurability of the Software-Defined Vehicle and the personalization enabled by the Flagship Experience is likely to result in many different interactions between the cabin's occupants and the vehicle. This will be particularly the case as customers transition from conventional to Software-Defined Vehicles. Intelligent simulators enable the automaker to model such interactions and determine whether they conform with the anticipated behaviors and desired user experience. An intelligent simulator is used with the appropriate digital twin. It provides a collaborative environment for the OEM and its partners. Conventional simulators are not new to the automotive industry. They have been used by automakers and their suppliers to understand the performance of certain vehicle electromechanical systems as well as vehicle electronics. However, simulators that incorporate AI capabilities have only been recently introduced to the automotive industry.

The data resulting from the simulations can be combined with data collected from the actual physical system to determine where the model and the actual system need to be refined or corrected. For example, a simulator can be used to predict and understand the Software-Defined Vehicle's ADAS response when pedestrians and bicyclists unexpectedly cut in front of it. The analysis of these responses may lead the OEM to develop and incorporate different sensors into the vehicle. Data from simulations are also used to develop machine learning models.

When companies started spending significant resources on the development of autonomous vehicles around 2015, where software and certain digital components were key for the vehicle's performance, a new breed of simulators emerged. These simulators borrowed many concepts from video games. They can work with very large data sets, use extremely detailed digital twins, and incorporate intelligent capabilities that allow them to identify problems and conditions that their operators could miss. For example, data depicting millions of photorealistic scenarios can be input into the simulators used to test a vehicle's autonomous driving performance, allowing the vehicle's designers to test every possible driving condition. Some of the input data may be generated manually, but most is synthetic and generated by other intelligent systems. The deployment of these simulators requires expertise in highly distributed systems, big data and AI, and even virtual worlds. The initial batch of these simulators that are coupled with digital twins have been developed by technology companies like Google and NVIDIA but also several startups like Applied Intuition, Cognata, and Foretellix. Incumbent automakers such as GM, Toyota, Audi, and Volvo license the simulators developed by startups.

5.3.3 Machine Learning Stack

AI and particularly machine learning play an important role in new mobility, Software-Defined Vehicles, and the Flagship Experience. For example, the Software-Defined Vehicle's infotainment system includes machine learning models for natural language understanding. But the machine learning techniques used to enable the Software-Defined Vehicle's automated driving or predict its battery charging needs are different from those used for natural language understanding. A machine learning stack must include components for data ingestion, data engineering, data science, and operation (MLOps).

5.3.4 Software and AI Models Version Control System

The software and models version control system stores every version of the software developed by the automaker for the Software-Defined Vehicle and each created machine learning model. While version control systems proliferate, very few organizations use systems for managing the versions of the various machine learning models they develop. Such a system makes it easy to understand each model's provenance, evolution through software transformations, the applications that use it, its resource requirements for training and inference, its deployment history, usage, and performance metrics, etc. The version control system stores the data needed for the development and updating of each machine learning model. It incorporates the data transformations performed on each data set, e.g., anonymization of personal data, cleaning of noisy data coming from vehicles, etc.

Because the vehicle's Software Platform is so integral to the Flagship Experience, the version control system must serve two goals. First, manage all versions of the vehicles' Software Platform supported by the OEM. As we see with consumer devices today, not all owners choose to use an application's or operating system's latest version. Contrary to what is happening with consumer devices today, when only the vendor determines how many of the older software versions will continue to be supported and for how long, because automotive is a regulated industry, the set of supported versions will likely be decided between the OEMs and governments. This is an issue that will have to be addressed by the automakers in collaboration with governments. Global automakers may be required to support a different set of versions in each geography they sell their vehicles.

Second, manage all versions of the machine learning models incorporated into the Flagship Experience services. New models are developed and maintained based on collected data. Maintaining model versions is different than maintaining software versions. In addition to the model's structure—typically captured by the model's architecture, e.g., the layers and nodes of a neural network and the connections between these nodes—it is important to incorporate with each version the data that was used to create the model, the data that was used to test its predictive power, and metrics relating to its ongoing performance.

The version control system provides the genealogy of every software module and every data element used in a data set. The genealogy facilitates software and data reuse and enables the OEM to better track costs and understand the configuration of each operating vehicle. This is also important because a data set will need to be transformed to fit the needs of each group that uses it.

5.3.5 IDEs for Vehicle Applications and Services

When the innovations introduced by iOS, iPhone's operating system, were understood and appreciated, an entire global ecosystem of mobile application developers was formed. Later, with the introduction of Android-based smart devices, the mobile application developer ecosystem grew significantly. The developers monetized their skills, while Apple and Google were able to enter new markets through the applications, keep their platforms fresh, and create a new business model and associated revenue stream. To support the mobile application developers, Apple and Google not only offered a broad distribution channel but also powerful tools to support the development process. At the core of these tools were the Integrated Development Environments (IDEs) offered by these two companies. Today Apple's XCode IDE is in version 14.2, whereas Google's Android Studio is in its ninth version, called Electric Eel. In addition to these IDEs, several companies are offering third-party IDEs, including several low-code and no-code variants that enable business analysts to easily develop, test, and deploy mobile applications without the need to know a particular programming language.

The combination of Software-Defined Vehicle and Flagship Experience incorporates two types of applications. Applications that are necessary for the vehicle's operation, which have been the focus of this chapter, and applications that are part of the Flagship Experience, as will be discussed in Chapter 6. Some of the latter run on the vehicle, some on the user's smart device, and others in the cloud and are accessed through a browser. The mobile application developers that will develop software to be used through the customer's smart device can use the existing IDEs to create and update their applications. However, through these IDEs, they will need to access the APIs the automaker wants to make public, as well as the software tooling APIs. For example, consider what is necessary to develop an application that predicts when the Software-Defined Vehicle needs to be charged to minimize disruptions to the driver's schedule. The software that implements this application will need to make a series of API calls and use the returned data to make certain decisions. In particular, it will call the API that provides the battery's charge level and the API that provides vehicle telematics data. The telematics data can be used by a predictive model to infer the driver's driving style. By combining the inferred driving style with the battery charge level data and data about weather and road conditions between the vehicle's current position and its destination and using a different AI model, the application can predict the vehicle's range. Once the range is established, the application can access data about the location and availability of charging stations to provide the driver with charging options. By having API access to this data and AI models, this service's developer can easily build the application. The application developed by the company CarIQ can be viewed as a service provided through the Flagship Experience. The application takes advantage of a variety of APIs to enable frictionless payments from within the vehicle to charging stations and parking locations.

The Vehicle Application IDE must enable the developer to distinguish between three distinct contexts that occur during the Software-Defined Vehicle's customer journey: as a transportation modality; as a space for the driver and the passengers (both separately and as a group) to socialize and even be entertained; and as an object that operates in a larger environment, e.g., a fleet or a city. For example, the Honda/Sony vehicle is built around the view that the vehicle is a mobile multimedia entertainment space. For each context, the developer must have access to different types of data. The telematics data generated by the Software-Defined Vehicle's sensors are important in the context of the vehicle as a transportation modality. It can be used to make inferences about the vehicle's performance, an important metric for keeping a fleet's uptime high, or predict the vehicle's range. Data collected from the in-cabin sensors, such as the occupants' infotainment choices and seat settings, including position and heating, are important in the context of the vehicle as a space for the driver and the passengers. Data about the vehicle's operating environment, e.g., other vehicles, pedestrians, road conditions, and the risk level associated with each one, are important in the context of the vehicle as an object in a larger transportation environment.

Automakers will need to choose whether to offer Vehicle Application IDEs to be used for the development of the applications and services that will run on top of the Software-Defined Vehicle's

Software Platform. These IDEs may be provided by the OEM.[8] Some Tier 1 suppliers have adopted the Apache Eclipse IDE (Eclipse Foundation, 2022). OEMs may decide to use it as well. IDEs may also be offered by the provider of the Real-Time Operating System (consider Blackberry's QNX Momentics IDE) or even the provider of the Infotainment Operating System. By providing the Vehicle Application IDE, the OEM will be in the position to incorporate in it the Flagship Experience software philosophy, tightly couple the Software-Defined Vehicle's architecture, and provide access to the OEM's software tooling, including the digital twins, intelligent simulators, data, and machine learning models. Digital twins and intelligent simulators can be used by the vehicle application developer to test each vehicle application's functionality before it is deployed as part of the Flagship Experience. Access to the data and the machine learning models enables the vehicle application developer to take advantage of these resources to enhance the application. By unifying software, data, security, and lower-level APIs into the Flagship Experience API, the OEM provides the developer with the ability to easily take advantage through a single API of the Software-Defined Vehicle's various platforms, as well as the data and software residing in the OEM Cloud.

By distinguishing between contexts, the developer will be able to determine an application's scope, e.g., how to enhance the driver's safety and incorporate the appropriate features. With each application, the developer will be able to establish the application's use and associated functionality but will also establish the connection between the application and the appropriate Customer Journey touchpoint. The supported versions of each vehicle application are maintained Configuration Manager described above.

8 Their development and support may be outsourced to a third-party software company.

The Software-Defined Vehicle is more than the sum of its parts. It is the nexus of the Flagship Experience, to which all of its software and related processes are essential, from the Software Platform in the vehicle's architecture to the apps that make up the Flagship Experience and the software tooling that enable its overall development. AI is found in many of these applications and platforms, increasing the utility and usability of each component. It is also at the core of the Customer Management Platform that will be discussed in the next chapter.

CHAPTER 6

Software for the Flagship Experience

During GM's Investor Day in October 2021, Mary Barra, the company's CEO, announced that by the end of the decade, the company will be generating annually $20 to $25 billion in new revenue from services offered to 30 million connected vehicles. During the same period, Stellantis expects to generate €20 billion from software and services to its 34 million connected vehicles, while Ford is forecasting $1,500 annually from each Software-Connected Vehicle using its version of ADAS and another $2,000 from insurance policies to such vehicles. Ms. Barra further claimed that $6 billion would come from the company's recently announced insurance offering. According to the company's market research, consumers are willing to pay up to $135 per month for vehicle-related services. As a measure of comparison, today's OnStar service generates $2 billion annually from approximately 4.2 million subscribers and 16 million connected vehicles, or about $40 per subscriber per month. If we assume that by 2030 50 percent of the eligible connected vehicle owners will pay for GM services, then the monthly revenue per user, excluding insurance, will range from $78 to $106. In other words, to meet its CEO's goals, GM will need to grow the average per-subscriber monthly revenue two to 2.7 times over the life of the vehicle. And it is not guaranteed that the entire monthly revenue will come from subscriptions.

Deriving revenue from the customers of Software-Defined Vehicles to achieve future financial goals will require OEMs to become customer-centric. Will offering the Flagship Experience entice the customer to acquire a Software-Defined Vehicle? Will the Flagship Experience with the customer understanding it enables and the personalization it supports increase customer monetization and loyalty? The answer to the first question depends on how each OEM decides to implement the Flagship Experience. The answer to the second relies on how good of an understanding is achieved and how well the OEM can match the value it offers to each customer's needs.

According to the automotive industry, when prospective customers (businesses or consumers) seek a new vehicle, they consider styling, performance (including its safety), reliability, fuel economy (or range, in the case of electric vehicles), TCO, financing terms, residual value, and technology. Judging from the recent increase in electric vehicle sales, they have also started considering the vehicle's environmental impact. Going forward, the customer must also understand and appreciate the key characteristics of Zone-Based Software-Defined Vehicles and the advantages they introduce over conventional and Domain-Based Software-Defined Vehicles.

Properly introducing the Software-Defined Vehicle to the market is a category-creation effort for the OEMs. The differences between Software-Defined Vehicles and conventional light-duty vehicles are analogous to those between "feature phones" and smartphones. They are both phones, but their use cases and values are vastly different because, through their customer experience and the large number of applications smartphones support, they enable their user to mobilize their entire digital life. Similarly, Zone-Based Software-Defined Vehicles through the Flagship Experience and their software can change ground mobility in ways that conventional vehicles, or even Domain-Based Software-Defined Vehicles, cannot. As it is being introduced, a new category needs to be anchored in something customers are already familiar with. For example, when Apple introduced the iPhone, it focused on three features to characterize the device. First, it was a phone. Second, it is a fully functioning web browser. Third, it is an iPod. There was no discussion about applications and the app store. These three accessible and understandable features made it possible for people to embrace the iPhone and then discover all its other category-defining benefits. If electric powertrains and OTA updates are two such key features defining the Software-Defined Vehicle over any other type of electric vehicle, the Flagship Experience's value-adding personalization can be the third.

The development and deployment of the Flagship Experience requires three platforms. The SDV Services Marketplace manages the services that are offered as part of the Flagship Experience. The Vehicle Management Platform manages each Software-Defined Vehicle's journey as defined in Chapter 3. The Customer Management Platform manages each end-to-end customer journey as defined in Chapter 3. These platforms run on the OEM Cloud but interact with and take advantage of the Software-Defined Vehicle's Software Platform.

6.1 The SDV Services Marketplace

Over the years, as smart devices have proliferated, consumers have become used to app stores as the places where they find the applications that extend the functionality of their devices to address the needs of their digital lives. As the number of mobility-related services offered as part of the Flagship Experience increases, it will be necessary for the OEM to create the SDV Services Marketplace. This marketplace will include applications and services that can be incorporated into the Software-Defined Vehicle, e.g., an application for parking and charging, as was described in Chapter 3, or the owner's smart device, e.g., an application for summoning a robotaxi service offered by the OEM. Some of these will be owned and operated by the OEM, and others will be offered by its partners. The OEM needs to identify the owned and operated services where it will have a competitive advantage and can excel. A good starting point is with services that are familiar and important to customers, e.g., auto insurance. With each such service, the OEM must not only incorporate it in the SDV Services Marketplace but work to educate the customers about the advantages the Software-Defined Vehicle brings to the service. The SDV Services Marketplace is based on a Services-Oriented Architecture that enables a service to publish the needs it addresses and the customer to subscribe to it.

Third-party applications may be developed specifically for the automaker's Flagship Experience or may be popular applications, e.g., Spotify, originally developed for other devices. To provide the right experience, these applications will have to be created, or reformatted in the case of applications originally developed for other devices, to take advantage of the vehicle's computing power and configuration, e.g., its monitors or its speakers. For example, gaming applications running on Afeela's vehicle will be reformatted to take advantage of its vehicle's monitors and speakers.

The SDV Services Marketplace may be developed by the OEM, as is the case with the VW Group, whose app store will be available in the VW, Porsche, and Audi Software-Defined Vehicles;

licensed from third parties, as is the case with Google's app store and Visteon's app store; or co-developed by the OEM and a third party, as is the case between Mercedes and Faurecia (George, "Volkswagen, Audi, and Porsche are getting their own in-car app store—and yes, that includes TikTok," 2023; Aftermarket News, 2022; and Guillaume G., 2022). The SDV Services Marketplace interfaces with the Vehicle Management Platform and the Customer Management Platform.

The SDV Services Marketplace provider must offer quality assurance services to validate that each submitted app offers the expected user experience, taking into consideration the vehicle's characteristics and configuration, since these will not only vary across models but also within the vehicles of the same model. This is why the App Store must interface with the Vehicle Management Platform. The quality assurance will also check the conditions under which each application can be used so that it will not endanger the safety of the vehicle's occupants, as well as the countries, or even the geofences, where the application can be used. A video conferencing application may not be invoked by the driver while he is operating the vehicle, while a given country may restrict the use of a social media application.

6.2 The Vehicle Management Platform

The Vehicle Management Platform manages the cradle-to-grave vehicle journey of each Software-Defined Vehicle in the OEM's Parc. It is an essential software component for the delivery of the Flagship Experience. The Vehicle Management Platform performs four tasks. First, it manages each vehicle's configuration, essentially each vehicle's digital twin (see 5.3.1). The configuration includes the components comprising the Hardware Platform and the Software Platform, as well as the applications and services running on the vehicle. A vehicle's configuration is modified as a result of the customer adding or removing services, OEM- or partner-initiated updates to the Software Platform, and other changes made to the vehicle, such as repairs requiring a visit to a service center. Second, it receives and manages the data generated by the Software-Defined Vehicle during its operation. Third, it manages the data associated with the vehicle's servicing and repairs. By interfacing with the Customer Management Platform, it links each vehicle's configuration and state with the data collected during the customer journey, as described in Chapter 3, and is associated with the customer. Fourth, it uses AI to make recommendations to the OEM about how to change the Software-Defined Vehicle's configuration after the vehicle is returned by its current owner and as it is being prepared to be reintroduced into the market for acquisition.

The Vehicle Management Platform includes a database, a configuration manager, the OTA update tool, and an AI analysis and recommendation application. The Vehicle Management Platform's database is the system of record of all the data relating to the Software-Defined Vehicles in the automaker's Parc. The key data structure is the *vehicle profile.* Each vehicle has a separate profile that stores all the data relating to each vehicle's up-to-date configuration, updates, repairs, etc., as it proceeds through the Vehicle Journey. The vehicle profile is connected with the profile of the vehicle's current owner (explained in the next section). The database utilizes the appropriate data pipelines to ingest and manage the data coming from each vehicle. Because OEMs sell their Software-Defined Vehicles in multiple countries, this database may be a distributed system.

The configuration manager provides a scalable way to manage the configuration of each Software-Defined Vehicle in the OEM's Parc. The configuration manager can be used as early as when a prospect starts determining what options and services to include in the vehicle under consideration, i.e., the pre-sales phase of the Customer Journey. Once the vehicle is delivered to the customer, the configuration is updated every time the customer adds a service through the SDV Services Marketplace or removes a service that was previously installed, when an OTA update is completed or because of an on-site service action.

Keeping track of each vehicle's exact and up-to-date configuration and the dependencies and averse interactions among the components is important for many different applications, including diagnosing the cause of a predicted problem and enabling OTA updates. Every time an OTA update needs to be pushed to a vehicle, the configuration manager and the OTA updater are invoked. The update may relate to the Software-Defined Vehicle's Software Platform or one or more of the applications running on the vehicle. Before updating an application running on the vehicle's Software Platform, the OTA update component checks whether the software to be updated is compatible with the Software Platform version running on the vehicle. If necessary, it first updates the vehicle's Software Platform before updating any application.

The OTA update component includes logic to determine when each update can be performed so that it will not endanger the vehicle's occupants. It may also include logic to minimize the communication costs associated with each update. Every time a software update is completed, the automaker must also guarantee that the vehicle's cybersecurity is not compromised and that the updated software will continue to function correctly with its current processors. As was mentioned in Chapter 4, cybersecurity needs to be designed into the software from the beginning rather than be approached after the fact; the automaker should have created a cybersecurity model by the time the development of the vehicle's software starts, and this model must be consistent with the model used by the vehicle's software architecture.

To install the correct update in each vehicle, the automaker must monitor and manage thousands, if not millions, of vehicle configurations within each geography it operates since each customer personalizes their vehicle differently. The vehicles in a fleet (or the Parc) may be running on a different version of the OEM's Software Platform and have a different set of applications and services installed. Very much like it happens today with our devices, certain versions of the Software Platform may be incompatible with versions of specific applications. The configuration manager must keep track of these combinations and their compatibility.

The Vehicle Management Platform's AI component (AI System 2 in Figure 3-2) analyzes the configurations of a user-selected group of related previously owned vehicles that must be prepared and made available for sale, utilizes how each vehicle's configuration evolved while it was operated by its previous owner, correlates them with indicators that can impact each vehicle's acquisition—e.g., interest rates, market demand, state of the supply chain, competitor offers, etc.—and recommends how to prepare each vehicle before putting it back on the market and how to price it. Preparing a vehicle may involve changes to its software and/or hardware configuration, e.g., adding, removing, or upgrading services, sensors, etc. The recommendations must consider the characteristics of the customer segment the OEM will target for the used vehicle. This operation is analogous to a realtor using market knowledge to recommend the changes a seller should make to their house before putting it on the market for sale. Over time the OEM should be able to determine whether the vehicle should be reconfigured, overhauled, or taken out of circulation.

6.3 The Customer Management Platform

Customer-centric corporations establish direct-to-consumer relationships and use them to collect, organize, and exploit data from customers, prospects, and devices (where appropriate). By doing so, they understand their behaviors and needs. They can communicate their value propositions on how to address these needs. Customer-centric thinking is not yet part of the incumbent automakers' corporate culture. However, automakers can take advantage of the experiences in these industries to expedite their transition to customer-centricity. They will need to become *consumer-facing, direct-to-consumer brands rather than manufacturers.*

Customer data collection is a prerequisite to customer-centricity, but it does not automatically make the organization customer-centric. Our work has shown that today the connected vehicle divisions of some incumbent automakers collect data from such vehicles. The marketing organizations of these automakers collect data during the pre-sales phase of the customer journey, e.g., a consumer's web searches about a vehicle model. But once the vehicle phase is completed, data collection during the ownership period is sparse.

In the same way that the Vehicle Management Platform is the key software for the vehicle journey, enabling the OEM to have a direct and continuous connection with the Software-Defined Vehicle, the Customer Management Platform is the key software for the customer journey. The role of the Customer Management Platform is to enable the OEM, as the owner of the customer relationship, to maintain direct and continuous interaction with each customer and prospect. The Customer Management Platform should include the following components, shown in Figure 6-1: a Customer Data Repository, a Customer Analytics component, a Customer Success application, the Recommendation Engine, and a Customer Relationship Management application. The Platform is part of the OEM Cloud.

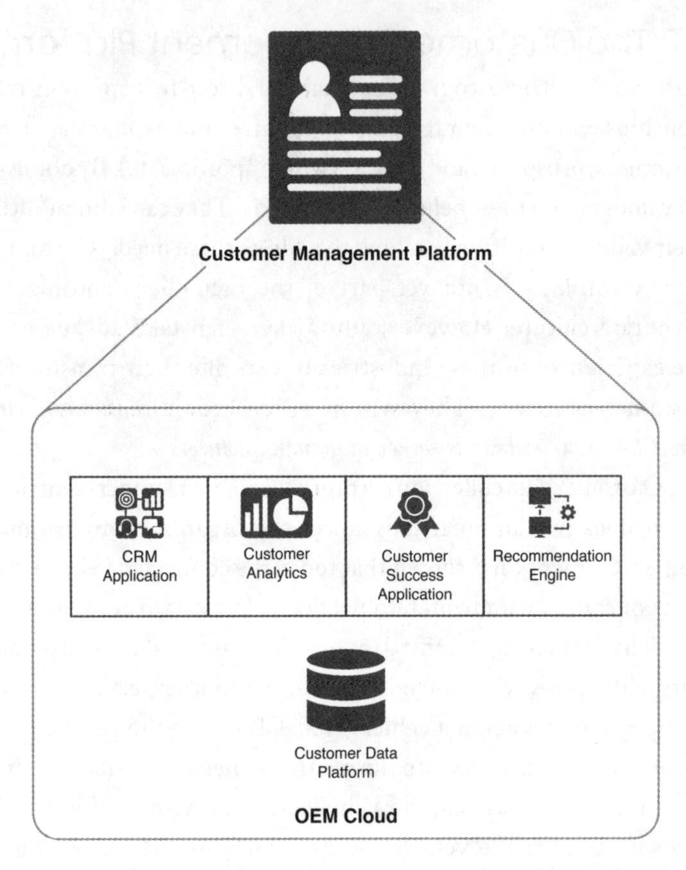

Figure 6-1: *The Customer Management Platform*

The Customer Data Repository organizes and manages each customer's and prospect's data that is collected from all the touchpoints regardless of the channel through which a customer interacts with the automaker (public and proprietary applications, websites, social networks, etc.) and makes it available to other components. The key data structure is the *customer profile*. Each profile is a graph that brings together and interconnects the variety of data captured about the customer's mobility. A profile is constantly updated with the data captured during the customer journey. With each customer profile, the OEM needs to associate a digital identity and a digital wallet. The OEM will need to determine how to connect, or otherwise relate, the digital identity and digital wallet with other identities and wallets the customer will likely have. In this respect, the OEM will need to determine a) whether to take on the responsibility of owning these, both of which have implications for the level of privacy and cybersecurity measures the OEM will need to establish, and b) how to instill in the customer the same level of trust that technology companies, such as Apple, have been able to establish with their customers. Because the data relating to the Flagship Experience is primarily personal data, the Customer Data Platform must include all the corporate and country privacy policies. Therefore, the data platform of an OEM that sells Software-Defined Vehicles in the EU and California must incorporate the GDPR policies for the EU and the CCPA policies for California (Haskins, 2022).

Customer analytics are used to analyze the data stored in the data repository to develop the deep customer understanding required for customer-centricity, create predictions about future behavior, and make recommendations based on the derived understanding and generated predictions. As we will see in the next section, the analytics component blends customer analytics with mobility analytics to generate relevant predictions and recommendations based on past behavior.

The Customer Success application focuses on the post-sales part of each customer's journey. It provides information on how a customer uses a product, as well as tries to explain why a customer may not use a product to its full potential. This information is used as input to the recommendation engine, which also has knowledge about the applications/services included in the SDV Services Marketplace and determines what action or service will be recommended next to each customer.

Incumbent automakers have rolled out customer relationship management (CRM) applications to use the customer data they capture. Dealers, which over time have consolidated into large corporations representing many brands, have deployed their own CRM systems. Their objectives to manage the customer relationship are often in conflict. The dealer wants to keep the customer as long as they acquire a vehicle from one of the brands they represent. For example, AutoNation does not care if the customer buys a GM, Ford, Toyota, or Nissan vehicle since these are all brands they represent. The objective of their relationship with the customer is very different from GM's or Ford's objective. As is incorporated into the Customer Management Platform, the CRM component becomes an interaction manager.

AI is a key technology used in the Customer Management Platform. The AI System 1 of Figure 3-1 is the Customer Management Platform. For example, predictive AI models can be used to flag inconsistent data in the Customer Data Repository or identify the individuals that belong to the same household and must be treated as such. It can be applied by the Customer Success component to interpret a driver's sentiment based on the interactions with the vehicle's model-defining or personal services. AI is at the core of any Recommendation Engine. The AI application implementing the FEAT framework, described in the next section, becomes the backbone of the Customer Management Platform. It combines customer and mobility analytics with the Recommendation Engine and much of the functionality performed today by a Customer Success component. Automakers will need to invest to either develop or license the software components that comprise the Customer Management Platform. They will need to have this platform in place by the time they introduce their version of the Flagship Experience with their clean sheet Software-Defined Vehicles.

6.4 The FEAT Framework

How OEMs analyze the data they collect and use the generated insights to provide value will determine the customers they acquire, retain, and monetize. To understand the customer as prescribed by the customer-centric strategy, enable the value exchange, and facilitate the personalization of the experience, we have developed the Flagship Experience AI Technology (FEAT) framework and implemented it as a data-driven AI application. This application is part of the Customer Management Platform. The FEAT framework and associated application provide customer and mobility analytics and match discovered characteristics with the value-adding services that are part of the Flagship Experience. FEAT combines customer understanding, customer needs prediction, and recommendations that turn needs into potentially monetizable service opportunities.

To accomplish these goals, FEAT utilizes four types of data. First, data about every trip the customer takes with the vehicle. When possible, it is also valuable to subsequent analyses to capture the data relating to all the ground trips the customer takes regardless of the modality used because this data reveals customer propensities toward specific modalities. Second, data about mobility-related spending. This includes spending on vehicle charging, parking, repairs, insurance, etc. Third, data about vehicle-related visits. For example, visits to the dealer or the OEM's service centers and any reviews associated with these visits. Finally, the data that is coming from the vehicle's in-cabin behaviors, e.g., the data generated from the interaction with the vehicle's navigation system. The driver's behavior is derived from the telematics data generated by the sensors of the Software-Defined Vehicle. If the Software-Defined Vehicle includes a driver-facing camera (typically included in vehicles that support some level of driving automation), the data describes the driver's condition during the trip, e.g., alert or distracted. These four data types are appropriately linked to create an intricate network. For example, a visit to a service center, which is a part of even generic customer journeys, to replace a component that is covered by the vehicle's warranty may be described as a simple entry of which component was replaced or may include the route the customer took, any intermediate stops he made on the way, and whether he waited for the vehicle to be repaired or departed the service center using a loaner vehicle or ride-hailing. In this way, mobility behavior is linked with the visit history data and the spending history data.

To capture mobility behavior, as early as 2020 Simoudis introduced the Urban Mobility Metric (Simoudis, The Urban Mobility Metric, 2020) that utilizes data from the vehicle, mobile devices, and urban transportation infrastructure to capture the trips a traveler, in this case, the customer, makes during a specific period, e.g., a month, within a geofence (typically a metropolitan area) regardless of the modality used. Each trip includes the places visited and the route that was selected to reach each place. For example, a daily commute from home to work may involve walking, riding a local train, and using a shared electric scooter, i.e., an on-demand micromobility service, to the destination. The Urban Mobility Metric also captures various aggregations and correlations of the trip data. For example, an aggregation of the total amount spent on each transportation modality during the period of interest. Or a correlation between a traveler's monthly mobility pattern and the city's mobility patterns captured through its transportation infrastructure, a correlation between the distance traveled using each modality to the weather conditions at each trip's origin.

FEAT enables its user to understand the automaker's customer base at the individual level. For each customer and prospect, it creates a profile. A customer profile consists of *characteristics, needs, services*, and a *customer journey*. FEAT populates the characteristics with data supplied by the prospect, relevant data in the OEM's Customer Data Platform, and data it infers. As additional data is collected from the prospect, the application refines the profile. A customer journey may be defined for each customer profile. Alternatively, the same customer journey may be associated with a set of profiles. Also, a portion of the customer journey may be the same for all profiles. For example, for some OEMs, the Flagship Experience's pre-sale process may be the same regardless of the brand or model, or even the prospect's characteristics. Tesla has the same pre-sales experience regardless of whether the prospect considers a Model 3 or a Model S Plaid. Others may choose to tailor that part of the Flagship Experience to each brand and model class they offer. The customer journey is not a static structure. As a result of the analyses performed using FEAT, the OEM may decide to update/extend a customer journey, refine the journey's touchpoints, change the existing recommendations, or add new ones.

When the prospect acquires a new vehicle, thus becoming a customer, three actions relating to the customer profile take place. First, additional characteristics including the brand-based, model-based, and personal services that comprise the customer-specific Flagship Experience are incorporated into the customer's profile. The FEAT application incrementally blends the four types of collected data and continuously analyzes it. In this way, it acquires additional customer characteristics and refines previously captured ones. FEAT's user can create a label for a customer based on the characteristics, e.g., "aggressive driver, road warrior, uses freeways, avoids congested routes, utilizes restaurant delivery services, shops at organic grocery stores, affluent, tech-savvy, environmentalist." Second, the configuration of the acquired vehicle is added to the Vehicle Management Platform. Third, the customer journey associated with the customer profile is further extended to incorporate the touchpoints associated with the Flagship Experience offered around the acquired Software-Defined Vehicle.

Each of the brand-defining, model-defining, and personal services (and features) in the SDV Services Marketplace are described in the FEAT framework. A service is described in terms of the *predicted mobility-related customer needs* it addresses, the *inputs* it receives, the *outputs* it generates, and the *attributes* of its output. Each output attribute addresses a *need*. How such services are implemented is outside the scope of this book. For example, assume that the OEM-owned personal service called "Mobility Insights" addresses four needs and has two inputs and four output attributes. The four needs it addresses are "Time Savings," "Cost Savings," "Safety," and "Energy Efficiency." It receives the intended destination and the vehicle's current location as inputs and generates four routes. The attribute of the first generated route between the vehicle's location and the intended destination is that it is the fastest. The second generated route is the cheapest. The third is the safest, and the fourth is the most energy-efficient. Because some of the services impact the vehicle's performance and/or capabilities, to match customer needs to such services FEAT must utilize the vehicle's model so that through AI reasoning it can determine what is possible. This requires that FEAT has access to the same model used by the vehicle's digital twin. For such a model to be used by FEAT or any other AI system, it is necessary to be implemented as an ontology using a knowledge graph structure.

FEAT incorporates a set of AI models that use customer characteristics to predict a customer's needs. For example, using the characteristics "affluent, tech-savvy road warrior, uses freeways, avoids congested routes," FEAT's model predicts that the customer will have two needs: a) the need called "Time Savings," and b) the need called "Cost Savings." A service is associated with an instrumented touchpoint. This association is created by the OEM since it owns the customer journey. A touchpoint may have several services associated with it. When the customer accesses the instrumented touchpoint, and his predicted needs match those addressed by the service associated with the touchpoint, an offer is made to accept the service. If the service is monetizable, then the customer is offered to license the service using the business model associated with the service, e.g., one-time transaction, monthly subscription, etc. Assume that the service "Mobility Insights" is associated with a touchpoint in the Software-Defined Vehicle's infotainment system. It is invoked when the infotainment system's navigation component is accessed and is offered as a monthly subscription, which the customer accepts. Upon subscribing, the service is downloaded from the SDV Services Marketplace into the customer's profile.

Once the customer subscribes to the service, the "Mobility Insights" service is invoked. The traveler provides the intended destination while the vehicle's current location is automatically obtained from the vehicle's sensors. These two inputs are supplied to the "Mobility Insights" service that automatically determines the safest, cheapest, fastest, and most energy-efficient routes between the vehicle's current location and the desired destination. The information generated by the service is sent to the vehicle's navigation system, which displays it for the traveler to choose.

The implementation of the FEAT framework into an AI application results in a complex system. To be in the position to realize the full benefit of such a system, an OEM and its ecosystem partners must have reached a certain level of AI maturity. *The Big Data Opportunity in Our Driverless Future* presented three levels of AI corporate maturity (Simoudis, The Big Data Opportunity in Our Driverless Future, 2017). Each level requires specific types of tooling for data management, data analysis, data science, model deployment, and insights visualization. The companies that have reached the highest level of maturity:

- Develop intelligent systems based on a previously defined strategy.
- Have established processes for data integration and cleaning, as well as machine learning model development and deployment.
- Consistently invest to support their AI initiatives.
- Integrate new AI applications faster.
- Attract better talent with the appropriate expertise and experience.

With Zone-Based Software-Defined Vehicles slated to be introduced in the middle of the decade, OEMs and the partners that will create the Flagship Experience must work now to ensure that they are in the position to deploy the FEAT framework.

6.5 Adopting and Monetizing the Flagship Experience

FEAT enables automakers and their partners to match the value of each available service to a customer's needs and recommend the service to the customer through an offer. Let's assume that it does a good job with matching and recommendation. Monetizing a service requires convincing the customer that the service will provide value. Ongoing monetization implies that the customer finds the service useful over time. How should automakers and their partners test whether their broader customer base will be willing to pay for services? Particularly at a time when surveys show that customers do not see most of the services offered by OEMs as worth licensing or buying. As with every other product introduction, an initial indication of its success is provided by the response of the early adopter and the early majority segments. Consider that it took almost twenty years for the financial services industry to transition its consumer customers from physical transactions to frictionless, digital transactions. On their way to success, financial services organizations realized that they needed to actively involve customers in the development of these services. They tested each service extensively with early adopters of technology before releasing it more broadly to other customer segments. The opportunity to provide input and shape each service enticed the early adopters to become heavily involved with these financial services industry programs. By adopting a similar approach an automaker can work with early adopters to receive feedback on Flagship Experience services, particularly personal services, before releasing them more broadly. Tesla has been using this approach, and GM and Mercedes are starting to do the same. This approach significantly increases the odds of adopting the Flagship Experience and the monetizable services it includes.

What are the characteristics of the likely early adopters of Software-Defined Vehicles? The closest we can come to answering this question is by looking at research conducted by Experian regarding the early adopters of luxury and non-luxury electric vehicles, most of which are Software-Defined. The research showed that the early adopters of such vehicles are members of Generation X, Baby Boomers, and Millennials, in this order. The early adopters of non-luxury electric vehicles have an annual income of at least $50,000 but most likely over $100,000. The early adopters of luxury electric vehicles have an annual income of over $150,000. According to the analysis, most of these adopters are professionals working primarily in technology fields, are already driving upscale cars, belong to households with a strong awareness of renewable energy, live in affluent suburbs, primarily charge their vehicles at home instead of public charging stations, and are sophisticated digital denizens. Early commercial adopters of Software-Defined

Vehicles are fleet operators whose vehicles operate in urban settings, e.g., delivery vans, service vans and pickup trucks, taxis, etc., that don't need to travel more than two hundred to two hundred and fifty miles per day and whose business requires the incorporation of software applications into each vehicle, e.g., applications for driver monitoring, predictive vehicle maintenance, etc.

Data relating to Tesla owners provide further support to Experian's research. The consumer early adopters of the Tesla Model S and later the Model 3/Model Y were primarily urban dwellers who were making a statement. They were proclaiming that they were environmentally friendly and appreciated the futuristic but unostentatious styling of the Tesla vehicles. They were also declaring that they were early adopters of cutting-edge technology, in this case, software, computer hardware, and vehicle technology, and embracing the complete Tesla value proposition, which dealt with range anxiety through the company's owned charging stations. They appreciated the services offered because of the vehicles' technology and started raving about it while overlooking many of the defects those early vehicles had. This is not the case with the buyers of the VW ID.3/ID.4. As mid-priced hatchback and crossover SUVs, respectively, they appealed to specific environmentally friendly customer segments, but because their technology did not work as expected, these versions were not embraced by early adopters in the way the automaker expected (Schmidt, 2021).

Since Zone-Based Software-Defined Vehicles will not enter the market until the middle of the decade at the earliest, OEMs should immediately prioritize and engage in continuous experimentation with this group regarding services, business models, and pricing options. The members of this group are already becoming familiar with OTA updates and the advantages of driving automation. Moreover, this experimentation should include acquirers of new vehicles but also early adopters of used Software-Defined

Vehicles in the market today since, as was discussed in Chapter 3, the buyers of these vehicles will represent an important constituency for the ongoing monetization of a vehicle through the Flagship Experience.

Customer lifetime value is the present value of the business attributed to the customer during the entire life of his relationship with the company. Therefore, in establishing the price of a for-fee service and then offering it to customers whose needs match the service's output attributes, the service provider (automaker or partner) must determine whether it fits the customer's lifetime value. The vehicle's total cost of ownership plays a key role in this calculation, and the vehicle's price is the key component of the total ownership cost. Therefore, before making an offer for a for-fee service, the service provider must be able to determine whether the customer will be able to afford it. The group of early adopters mentioned above can afford vehicles that cost at least $75,000. But the adoption of Software-Defined Vehicles, including Zone-Based ones, by the broader market will require vehicles that are priced in the $25,000 to $35,000 range. It also remains to be seen whether households that can only afford one vehicle will choose one that is battery electric as the Software-Defined Vehicles are.

According to the National Automobile Dealers Association (NADA) 2022 report, the average new light vehicle price in the US in 2015 was $33,456, whereas in 2022 it was $46,287, an increase of 38.4 percent (NADA, 2022). GM has introduced the Chevrolet Equinox EV, a Software-Defined Electric Vehicle whose entry-level model will be priced at $30,000, and announced a collaboration with Honda to develop additional lower-priced Software-Defined Vehicles, but in general, incumbent OEMs are facing difficulties offering Software-Defined Electric Vehicles at this price range (Capparella, "Honda and GM Deepen Ties, Promise 'New Series' of Affordable EVs," 2022). The average price of

Chinese Software-Defined Electric Vehicles is closer to $35,000 (JATO Dynamics 2022). Chinese OEMs like BYD, Li Auto, and others are ahead in this area, each offering several models in this price range or below (Yan and Goh, 2023; Cushing, 2023). Because of what consumers in the US and Europe are used to paying for a vehicle, without vehicles in this price range, it is hard to think of market penetration at the 50 percent level necessary for the Flagship Experience's adoption beyond the luxury segment. Today an electric vehicle costs a little over $9,000 more than a comparable conventional vehicle. To lower the initial acquisition cost of the Software-Defined Electric Vehicle, OEMs can consider offering Zone-Based Software-Defined Electric Vehicle models with very basic configurations, which presumably will have a lower price, and allow the customers to customize and upgrade them post-sale by adding the desired features using OTA updates.

While early adopters of Software-Defined Vehicles may be monetized relatively easily because of their affluence and extensive digital life connections, the broader market must be guided to consider the TCO and the residual value of these vehicles rather than only the purchase price. This will help them determine how much they will be willing to spend to have access to the mobility experience that clean sheet Software-Defined Vehicles offer. In today's vehicle acquisition experience, the vehicle's characteristics are determined at the time of manufacturing with a few options incorporated by the dealer. This process impacts the vehicle's TCO since the vehicle's Manufacturer's Suggested Retail Price is heavily impacted by its built-in features. Several studies conducted over the last three years have demonstrated that the TCO of electric vehicles is lower than the TCO of similar ICE vehicles (Burnham, Gohlke, Rush, et al., 2021). These studies considered factors such as monthly payment, maintenance, and charging, but also insurance and residual value. For example, the average monthly payment for a 2021 Tesla Model 3 Standard Range Plus is $409 with a down payment of $5,604 on a three-year lease (Doll, 2021). According to a report by the Nickel Institute, the three-year North American TCO for a high-mileage Tesla Model 3 is $13,260, whereas the corresponding TCO for the Chevrolet Malibu is $21,675 (Nickel Institute, 2021). According to the same report, the three-year TCO for a high-mileage Tesla Model S in Germany is €11,127, whereas the corresponding TCO for the BMW 5-Series is €21,906. The big difference is likely the maintenance costs, which can be significant for ICE vehicles but nearly zero for battery-electric vehicles. The lower the TCO, the higher the amount that can potentially be devoted to the after-sale acquisition of features and services.

For the time being, beyond expressing their aspirational financial goals, incumbent automakers do not have concrete strategies to consistently monetize the customer over the life of the Software-Defined Vehicle, generate revenue streams that are less cyclical than their existing ones, and achieve profit margins comparable to those reported by companies in other customer-facing industries. The massive investments OEMs are making around Software-Defined Vehicles, combined with the changing automotive market dynamics due to geopolitics and the fluctuation in the price of important materials (e.g., minerals used in batteries) and components (e.g., advanced semiconductors) could mean that Software-Defined Vehicles will not be profitable at the time of sale. The monetization of post-sales services may become the only way for an automaker to generate profits around the Software-Defined Vehicle. Therefore the OEM must develop monetization strategies. These strategies should enable the automaker to achieve the promise of the Flagship Experience of monetizing each Software-Defined Vehicle during the entire Vehicle Journey and each customer over the length of the relationship they have with the OEM.

Below are four strategies that can achieve this goal in the process of monetizing the customer. Though they should not be considered exclusively, subscription models represent a viable candidate for monetizing elements of the Flagship Experience. Subscription models have been successfully employed by other industries in conjunction with customer experience implementations to improve loyalty and customer lifetime value and should be of interest to automakers. Companies that employ subscription-based revenue models use highly configurable products in combination with digital customer-facing platforms that enable them to use the set of strategies provided below.

Strategy 1: Invite each prospect to try out for a limited time the Software-Defined Vehicle, and the services offered around it, before signing up as a customer. During this evaluation period, the tester is given access to the product's complete feature set to help them decide how to configure their vehicle. Even though they will be using the product for free, they will receive the same support as every other customer. In this way, they experience both the service and the quality of its support. Because products offered through subscription-based business models are highly instrumented through the data they generate, the automaker can determine how the customer using the service responds, how frequently they use it, and which functionality they find harder to use. Such feedback not only informs the sales process but also results in important input to the product development and product support teams. In this case, it can impact both the service's development and support teams and the corresponding Software-Defined Vehicle teams. If the prospect would prefer to continue evaluating the product beyond the initially allotted period, the leading companies in the use of recurring revenue models can support for a period the extension using other models, including, oftentimes, advertising models, while continuing their efforts to convert the prospect to a subscriber.

Strategy 2: Allow the customers to license only the services they need and keep using them for as long as they need them. Leading products monetized through subscription-based models offer a base version with a minimal set of capabilities and a large menu of services that can be selected by the customer on demand. Automakers can offer each Software-Defined Vehicle model in two or three base trims and enable the customer to select from a catalog of applications to be dynamically installed on the selected trim via OTA software updates. The customer licenses these applications individually, in packages, or using an all-you-can-eat subscription, not unlike what is currently happening in cable television, where the consumer can license a package that includes specific channels, and streaming video services, where companies like Netflix employ the all-you-can-eat subscription model. It must be possible to vary the length of the subscription from a few months to multiple years. Cancellation of specific services or the entire product subscription is always available. While subscriptions should be the most prevalent business model used for licensing these additional services, transaction-based models may also be employed in specific cases.

Strategy 3: Continuously measure the customer's satisfaction level with the product using metrics such as the Net Promoter Score, Customer Churn Rate, and others. These metrics are typically combined with the contracted revenue (i.e., the value of the licensed features) to determine whether the customer's satisfaction level will result in a simple subscription renewal (i.e., the customer extends the subscription to the features they had already selected) or a larger subscription amount (i.e., an upsell) which may be the result of licensing additional product functionality.

Strategy 4: Continuously engage over the customer's preferred digital channels to understand the customer's needs and determine how to address them. By doing so, the customer feels that their voice is heard, and they continuously receive value from the automaker. As a result, their satisfaction and loyalty increase. The customer can immediately select from recommended services and add them to their subscription. The OEM can also proactively identify and address product issues before every customer encounters them and determine whether a specific customer, or an entire organization, needs additional training on the product before they can take full advantage of what they have licensed and offer it (often for an additional fee). Even if the customer decides not to license the new features, they are satisfied that the vendor is proactively and constantly striving to offer them additional value.

Incumbent automakers are experts in manufacturing and financing their conventional ICE vehicles. But they face significant risks and have a steep learning curve to climb as they introduce Software-Defined Vehicles and transition away from ICE vehicles. They must invest to transform the technology, manufacturing, distribution, sales, and support models and create new revenue streams by offering the right customer experience. The adoption of the Software-Defined Vehicle that is upgradeable and customizable over its entire lifecycle through the Flagship Experience opens new revenue opportunities to automakers but also harbors new challenges in each of these areas. Customers, be they consumers or businesses, will have to appreciate the value of the vehicle before attempting to appreciate and adopt any of the services the OEM matches to their needs and offers them. OEMs and their partners will need to ensure that each selected service is priced properly, offered with the business model that fits the intended customers and their lifetime value, is delivered consistently, performs as intended, and continues to provide value to the customer on an ongoing basis.

CHAPTER 7

Organizational and Business Model Transformations

To bring to market Software-Defined Vehicles and the Flagship Experience, automakers will need to assess their organizations in terms of people, process, technology, and business model. The automaker must evaluate the people they have and determine the personnel gaps they will need to fill, decide on which organizational structure will help them deliver vehicle and experience, optimizing cost and risk, develop or acquire the technologies they need, and agree on the business models they will adopt. Success will require not just one of these four elements but all four. Based on their decisions in these areas, OEMs can be organized into four categories. Chapter 5 described the technological transformations necessary for developing the software for Zone-Based Software-Defined Vehicles. In this chapter, we focus on organizational and business transformations. In addition to describing the categories and the organizational and business transformations they require, this chapter presents past transformation efforts by the incumbent OEMs, draws lessons from these experiences, and offers recommendations for successful outcomes.

7.1 The Four OEM Categories

In *Transportation Transformation* (Simoudis, 2020) I presented a framework to organize automakers into categories based on how they innovate to best operate in new mobility. This categorization enables an OEM to understand the transformations they needed to undertake to successfully transition from a category to a different target category. The framework measured two dimensions of innovation: technology and business. Using this framework we can organize incumbent automakers into four categories shown in Figure 7-1.

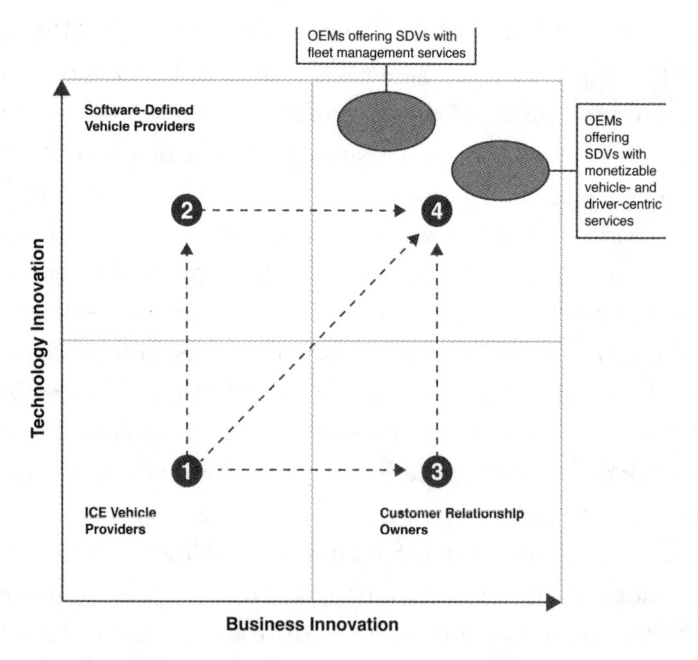

Figure 7-1: *Four OEM categories based on technology and business innovations*

The first category (Category 1) includes the automakers that continue to produce Internal Combustion Engine (ICE) vehicles, which today includes all the incumbents, and offer these vehicles exclusively through dealer networks.

The second category (Category 2) includes the incumbent OEMs that innovate along the technology dimension to produce Software-Defined Vehicles, e.g., Ford, VW, GM, etc., and offer them using a conventional customer experience through their dealer networks. We include in this category the incumbent automakers, e.g., GM, Toyota, BMW, and others that offer a few driver-centric services, e.g., GM's OnStar, with their newer ICE vehicles, because these vehicles are connected. However, these services are offered in the form of a catalog for the customer to choose from rather than in response to a perceived or stated customer need. Our research shows that GM and Mercedes have thought more deeply about what services to offer and how to monetize them than other OEMs in this category. Nonetheless, in these two categories, the dealer is the primary customer relationship owner.

The third category (Category 3) includes the incumbent automakers that innovate along the business dimension but not about Software-Defined Vehicles. These innovations are typically driven by the automakers' goal to derive significant revenue over the life of the customer relationship rather than maximizing their revenue opportunity at the time the vehicle is first sold. This approach involves business model innovation. For example, some automakers e.g., Mercedes, Porsche, and Volvo, offer vehicle subscription models (in addition to sales and leasing). This is innovative.

The fourth category (Category 4) includes the incumbent automakers that innovate along both dimensions. These are customer-centric automakers offering Software-Defined Vehicles that monetize the customer and the vehicle throughout the customer and vehicle journeys respectively. Over time, these OEMs may be able to derive close to 50 percent of their revenue from services, with the rest coming from the actual vehicle and replacement parts. Today Category 4 includes two OEM groups. The first group includes the OEMs that offer fleet-based services using Software-Defined Vehicles, such as GM's BrightDrop and Cruise and Ford's Ford Pro. The second group includes newcomer OEMs, e.g., Tesla, Rivian, Nio, etc., that offer monetizable post-sale services with their Software-Defined Vehicles. The incumbent OEMs that will offer Software-Defined Vehicles with the Flagship Experience will belong to a new group within Category 4.

The arrows in the figure indicate how by undertaking certain transformations an OEM can move to a different category and ultimately into Category 4. Moving from Category 1 to Category 2 requires the technology transformations described in Chapter 5 and organizational transformations described later in this chapter. As we will see below, organizational transformations may range from creating a new division to creating a new company. VW created a company called CARIAD to develop the software for its Software-Defined Vehicles, thereby moving from Category 1 to 2. A radical transformation to enable an incumbent automaker to move from Category 1 to Category 4 could be through major acquisitions of newcomer OEMs. This is a transformation that was performed successfully by large pharmaceutical companies and continues to this day. In particular, large pharmaceutical companies introduce new products and adopt new business models by acquiring private companies whose products are in advanced stages of approval by the US Food and Drug Administration. In this way, they forgo the risk associated with early-stage drug R&D while using the power of their distribution channels. Similarly, incumbent OEMs may acquire private companies that have already developed Zone-Based Software-Defined Vehicles and established direct-to-consumer models, e.g., Lucid Motors, and use their financial resources and established market presence to grow the acquired product.

These transformations are not always successful. A few years ago, to compete with independent app-based mobility services companies like Uber, Lyft, and Via, incumbent OEMs such as GM, Ford, Mercedes, and BMW undertook expensive Category 3 transformations. They created separate companies or acquired startups to offer on-demand ride-hailing, car-sharing, and microtransit services using their ICE vehicles. Mercedes offered ridesharing services through Car2Go, and GM offered car-sharing services using Maven, while Ford acquired Chariot to offer a microtransit service. However, most of these OEMs have since abandoned these efforts. One may say that these efforts failed because the OEMs were trying to imitate models that were not a good fit for their industry. But the failure also showed something about the industry's inability to take on a radical transformation of the type these models necessitated.

Moving to a different category does not mean that the OEM stops being a member of the category it starts from. An OEM may belong to multiple categories. For example, today all incumbents are members of Category 1, but GM also belongs to Categories 3 and 4 because it continues to make ICE vehicles, has introduced Software-Defined Vehicles, and provides fleet-based services (via BrightDrop and Cruise). Ford belongs to Categories 1, 2, and 4 because it has introduced Software-Defined Vehicles (Mach-E, and F-150 Lightning) and offers fleet management services (Ford Pro) for these vehicles. Over time incumbents will abandon Category 1, though their timelines differ. However, not all OEMs that have moved or are planning to move to Category 2 will move to Category 4. They will simply offer Software-Defined Vehicles using the existing business models and customer experience. On the other hand, as they abandon Category 1, those that have moved to Category 3 will automatically be abandoning it as well.

The transformations to first develop Software-Defined Electric Vehicles incumbent OEMs are undertaking are expensive, painful, and long-lasting. They are expensive because they require the automaker to make major R&D investments in the technologies Software-Defined Vehicles require. For the past several decades incumbent automakers minimized R&D investments by passing the burden of creating new automotive technologies to their suppliers. The industry reduced production costs by becoming the integrator of large systems provided by the suppliers and unit costs by signing long-term contracts with them. This approach leaves no room for the ongoing vehicle upgrading and experience personalization that is at the heart of what newcomer automakers offer and what the Flagship Experience is all about. Outsourcing product R&D may also make sense if you believe that your partner understands the market's changes and customers' expectations better than you. But in this case, not being customer-centric themselves, the partners didn't. OEMs missed important signals about the new mobility customer expectations. Rather than optimizing for cost, the newcomer OEMs optimized for customer experience and owning the customer relationship. These transformations are painful because they will require both internal organizational changes, but, more importantly, as the OEM transforms to become customer-centric they will require the OEM to establish a different relationship with the dealer.

The automaker has been locked on a specific business model, based on which the vehicle is monetized when it is shipped to the dealer rather than over the Vehicle Journey, with little or no opportunity to monetize each customer that owns the vehicle rather than just the first customer. The dealers optimize around the vehicle sale and maintenance and mostly for the period during which the vehicle is under warranty. Dealers that represent multiple brands optimize around keeping the customer without focusing on which of the brands they represent the customer will choose. The transformations will be long-lasting because of the industry's scale. No matter how nimble incumbents want to be, if they decide to transform to move exclusively to Category 4, or even just to Category 2, their decisions and associated investments will have a long-term effect.

7.2 Past OEM Transformation Effort

Transforming along one dimension can be hard. Successfully transforming across both the technology and business dimensions, as incumbents will have to do to enter Category 4, will require ambidexterity. To make matters harder, for a period the incumbent OEMs will need to operate both under their existing model of not owning the customer relationship and under the new model in which they will be the primary relationship owners.

To realize the Flagship Experience, organizational transformations are necessary to create corporate cultures with a vision for new mobility, and business transformations are necessary to facilitate the customer's monetization, particularly post-sales monetization. Both of these transformation types will be hard for an OEM to implement. As Stephan Durach, senior vice president of development at BMW's Connected Company, acknowledged to the author "These transformations are ongoing and could last several years." In 1995 Amazon started selling books over the Internet. Over the past thirty years, through technology and business transformations, it has redefined not only retailing, computing, and media, but also the customer experience we associate with each of these sectors. In the retailing industry alone, to compete effectively with Amazon, companies such as Walmart, Target, Kroger, and

Costco, had to undergo extensive and expensive technology and business model transformations that, in many cases, continue to this day. Today, in the automotive industry, Tesla plays the role Amazon played in retailing. It didn't just introduce a new car with a new powertrain. It brought ambitious and holistic thinking to an industry that for the past several decades it was only tweaking its practices.

This is not the first time that OEMs have attempted multifaceted transformations to address strategic challenges. It is instructive to consider in more detail two such past attempts: GM's Saturn Corporation and BMW's Project i. These two efforts were similar in both goals and execution to the transformations incumbents will need to successfully execute to transition to Category 4. They both involved the development of new vehicles, i.e., the Saturn S-Series and the BMW i3, that required technology transformations both for the vehicle and its manufacturing. In both cases, the vehicles were going to be offered under a new customer experience that was designed in parallel with the vehicle, rather than as an afterthought, and introduced new business models. They required major organizational transformations. In Saturn's case, GM created a separate company, not unlike what Ford did with Model E. Finally, in BMW's case, the customer's monetization beyond the sale of the vehicle was one of the main goals. While neither achieved its original goals, they provide important lessons for automakers that want to offer the Flagship Experience to better monetize the customers of their new Software-Defined Vehicles.

7.2.1 GM Saturn

GM conceived Saturn Corporation in the mid-eighties as a response to the threat posed by Japanese fuel-efficient and affordable cars. It was a "moonshot" initiative—the name refers to the Saturn V rocket that took men to the moon—by GM's CEO Roger

Smith, with the support of Donald Ephlin, at the time the president of the United Auto Workers (UAW) union. The goal was to create a line of vehicles that could compete with Japanese imports in the American market, which could be offered a new customer experience. Saturn Corporation was established as a standalone entity rather than a GM division. The first cars, the S-Series, were introduced in 1990. Technologically speaking, the S-Series was competitive with the Toyota Corolla but not better.

Among Saturn's corporate innovations, it ushered in new approaches to managing the workforce and fostering a more cooperative and customer-centric culture in the workplace. The employees were organized into work units, each with fifteen members. Each work unit was self-directed. Saturn's managers were free to build the plant without the usual corporate requirement of using a certain amount of existing tooling or standard parts on the car. They established a partnership with the union workers that was based on simple principles. Because vehicle quality was viewed as the key requirement for customer satisfaction, the workers could stop the vehicle assembly process to fix a problem. The company gave workers profit-sharing that was linked to the success of the Saturn cars instead of the entire corporation's performance. An educational process was established so that the union leaders could become more business-oriented and managers could understand how employees feel about the business.

Saturn introduced an important business model innovation through a no-negotiation sales price. To facilitate this, the company created a new dealer network, designed in such a way as to ensure that each dealer had a large territory so that the new business model would not negatively affect the dealer's economics. Saturn was thereby able to diverge from the typical dealer franchise agreement, whereby the dealer can set their price. Dealership employees were selected and trained to be customer-centric, collaborative, and accountable. The employee's goal was to help the customer select the right vehicle, not to negotiate with the customer on price. They had access to the entire production pipeline to expedite order changes and maximize customer satisfaction. The dealers were firm believers in the company's mission. For example, to address the growing demand in 1993, the dealers rebated back to the company 1 percent of each vehicle's sales price so that the automaker could start a third production shift (Hanna, 2010). It should be noted that the customer's post-sale monetization was not part of the target experience, as is in the case of the Flagship Experience.

Saturn created several technical innovations. The plant didn't use the moving assembly line that was the practice then. Instead, the cars were shuttled from station to station, where groups of workers attached "modules," or large finished sections of the car, while the vehicle was at rest. Stamping, machining, and casting were done within the Saturn plant instead of other locations, as was the case with other US automakers at the time. Saturns were plastic-bodied vehicles. This was possible because the company introduced large-scale injection molding. As a result, the vehicles were light, which added to their fuel efficiency. Their plastic bodies made them resistant to dents and rust. Rust resistance was particularly important for certain American markets such as the Northeast and the Midwest. But the company also missed, or refused to implement, other innovations the industry was adopting, the most important of which was the Toyota Production System.

The implementation of the innovations Saturn adopted required transforming how the workforce approached the design, production, and support of vehicles, which relied on a unique agreement between the automaker and the union. While Saturn hired employees with no prior GM affiliation, it also hired many employees from other GM divisions. These were selected because they were innovative risk-takers. The technology transformations, such as the manufacturing of plastic vehicle bodies, required the invention of new assembly line technologies and processes. The creation of these processes relied on hiring the right people, supporting them with the required level of investment (initially $5 billion), and not relying heavily on the use of components that already existed within other GM divisions, something that changed later.

Unfortunately, the transformations Saturn undertook were ultimately abandoned, and its innovations were overlooked by the rest of GM despite the automaker's significant investment. Among the reasons for Saturn's failure, we note GM's decision not to continue supporting it in the presence of mounting losses, lack of new wide-appeal models to follow in the success of the S-Series, in-fighting with other GM divisions that saw resources being diverted to Saturn, changes in GM's and the union's management, starting with Smith's and Ephlin's retirements, and resistance from unions to continue supporting the unique relationship between labor and automaker. The culture that had been so instrumental in Saturn's birth and initial success was no longer there by the end.

7.2.2 BMW Project i

BMW's Project i was set up in 2008. It was championed by Norbert Reithofer, BMW's CEO at the time. Its goal was to create and offer electrified (all-electric and hybrid electric) premium vehicles and associated services that addressed the needs of people living in megacities. In 2011 BMW created a new sub-brand, BMW i, to offer the vehicles developed under Project i. The first vehicle was the BMW battery electric i3, which entered the market in 2013. The i-brand introduced the i8 in 2014, the ix, and the i4 in 2020. BMW planned to offer a variety of services to the owners of i-brand vehicles. These services included parking, charging, multimodal journey planning, car-sharing, and ride-hailing. The first of these services, a car-sharing name under the brand DriveNow, was introduced in Europe in 2011 and expanded in the US under the brand ReachNow in 2016. By that year a ride-hailing service was also added. When introduced in 2011 DriveNow used only i3 vehicles.

In addition to the post-sale and mobility services, the i-brand pioneered several innovations. The i3 was a battery-electric vehicle with vertically integrated production: chassis, powertrain, batteries, and body were all designed and manufactured by BMW. The vehicles were produced in manufacturing plants that were using energy from renewable sources, had reduced CO2 emissions, and consumed significantly less water during production compared to the typical automotive manufacturing plant. The i3 had a carbon fiber body to help reduce its weight and incorporated renewable and recyclable materials in its interior. It was designed from the ground up to address the customer experience of driving in a mega city. This made the i3 a perfect fit for addressing certain of the new mobility needs discussed in Chapter 2. The vehicle's body shape, its doors, seat placement, battery range, and the materials used in its interior were all selected as part of the holistic customer experience design. The i3 also included the ConnectedDrive in-cabin services and provided access to one of the first vehicle-specific application stores. BMW was one of the first OEMs to offer a comprehensive set of mobility services that enabled the company to introduce a new revenue stream under a new business model.

BMW moved many employees from other divisions to the i-brand organization, including Ulrich Kranz, a thirty-year BMW veteran who initially headed the i-brand. BMW also made key external hires, such as Dieter May, who came from Nokia to lead the digital services organization, and Richard Kim, who was responsible for the i3's design. The introduction of mobility services around the i3 necessitated viewing the vehicle itself as a service rather than just as a product that can be owned. In addition to the tools BMW developed in the process of creating the i3, the technology transformations it initiated resulted in partnerships with several startups, like Ridecell, which provides fleet management software, and ChargePoint, which provides a charging network. Many of these companies had received investment from iVentures, the venture capital group that BMW set up as part of Project i.

In July 2022 the production of the i3 ended after a production run of 250,000. Similar to Saturn, BMW did not follow the i3's introduction with other wide-appeal models until recently. The i3 and later the i8 were targeting niche markets and were expensive to produce. It took almost ten years from the time the i3 was announced for the i-brand to introduce an SUV aimed at a broad market. As a result, today BMW is not viewed as a leader in the development of clean sheet Software-Defined Vehicles. Some of the mobility services introduced with the i3 have been discontinued, while others evolved into standalone companies. After a short period during which BMW and Mercedes merged certain of their mobility services into a single organization, they sold the unit to Deutsche Bahn. Project i and the i-brand didn't always have the support of the company's top management because of the i3's low initial sales compared to the large investment the sub-brand was receiving. Several Project i innovations have transferred from the i-brand to other BMW divisions and BMW Group brands. The 5-Series received the eDrive, and ConnectedDrive is now part of every BMW vehicle.

It is important to consider the i-brand, and particularly the i3, in the context of this book for two reasons. First, the design of the i3 was driven by a target customer experience. The customer experience, which included various post-sale services, as well as in-cabin services that were designed specifically to address the customer's urban mobility needs, was groundbreaking. While the vehicles were not sold in a differentiated way, the availability of mobility services that complemented private car ownership was innovative at a time when the importance of multimodal mobility was not appreciated by the average consumer. Second, launching Project i and the i-brand required BMW to create a new corporate structure and look outside the company to hire people with the appropriate experience and expertise to reduce the project's execution risk. It established new recurring revenue and transaction models that were deemed appropriate for the post-sales services it offered. The i-brand started creating a network of partners at a time when the automotive industry was working exclusively with hierarchical partnerships. The i-brand also used vertical integration (vehicle, batteries, charging network, and other mobility services), which was counter to the approach the rest of the company, or even the incumbent industry, was taking at the time.

7.2.3 Six Observations

Even though Saturn's S-Series and BMW's i3 were hailed as innovative vehicles, Saturn Corporation ultimately failed, and the i-brand continues only after making a major pivot and ceasing production of the i3 and i8 models. Because the goals of these two efforts relate to the goals that will be necessary for incumbent OEMs to achieve the move to Category 2 and potentially to Category 4 through the Flagship Experience, we can draw the following six observations from GM's and BMW's experiences.

Observation 1: ***The right organizational culture is a prerequisite for long-term success.*** I attribute the initial success of the Saturn Corporation and the i-brand to the cultures of innovation and risk-taking created within these organizations by their founding executives. These organizations were true startups. Employees were able to set their practices and optimize them as they saw appropriate. They were not encumbered by the norms and culture of their parent corporations. They could develop new technologies and establish partnerships with companies outside their parents' existing partnership set. Over time, the culture changed and reverted to that of the parent. This trend appeared to have accelerated once the executives who championed these organizations left the parent. This was also because more employees transferred from the parent organizations, who did not share the risk-taking and innovation characteristics of the founding employees. By transferring employees from other organizations, GM and BMW saved money, including money they would potentially have to allocate for severance packages to existing employees, and were assured to be working with "known entities" that were familiar with the parent company's culture. Establishing and maintaining the appropriate culture will be key to the success of the Flagship Experience.

Observation 2: ***Organizational resistance due to culture clashes can be a major inhibitor to success.*** For such large-scale efforts to succeed, everyone involved needed to be "all in." This was not the case with either Saturn or the i-brand. Both efforts faced stiff resistance and a lack of support from management, employees, and even partners of their parent organizations (Staff, 2009). My experience advising corporations on their innovation efforts leads me to conclude that this type of organizational resistance is typically attributed to the misguided belief that supporting the new organization will result in job insecurity and diminished budgets. The result of the established corporate culture entering the startup organization is the dilution of the original vision and cannibalization of the initially-devised plan, resulting in unachieved goals. By introducing components from other GM vehicles, the later Saturn models were completely undifferentiated and uncompetitive in the market.

Observation 3: ***Ongoing investments are necessary.*** Both the Saturn Corporation and the i-brand received generous initial investments. For example, in the nineties, GM invested $5 billion in Saturn, which would be $10 billion today. BMW initially invested over $2 billion dollars in Project i (Reed, MacDuffee, and Hrebiniak, 2021; Meiners and Knoedler, 2013). Most new ventures not only require startup funding but also require ongoing investment through years of unprofitability as they try to scale. Saturn and the i-brand were no exception. In both cases, GM and BMW decided to stop investing. Due to the lack of ongoing investment, Saturn didn't complete the transformations that would have made it the automaker Smith and Ephlin imagined. One can even say that the company stopped innovating after the S-Series was introduced. Similarly, it took a long time for BMW to restart investing in the i-brand and follow the i3 and i8 with new models. The lack of new models didn't widen the brands' appeal to new market segments and did not lead to the existing customers becoming repeat buyers. On the contrary, we saw Tesla continuing to invest along the entire value chain as it worked through many years as an unprofitable company. As was mentioned earlier in this book, automakers have committed large sums towards the development of Software-Defined Vehicles. These commitments were made while the global economy was in a multiyear growth phase. As the economy slows down, it will be important for automakers to understand the implications of diverting from their announced investment plans. While the Flagship Experience will provide automakers with new revenue streams, there is no question that the development of Software-Defined Vehicles will require large investments over several years.

Observation 4: ***Multifaceted transformations require the right skillsets.*** The success of transformations depends on bringing new skillsets and integrating them under the adopted organizational culture. This did not prove sufficient. Both Saturn and the i-brand hired people from other companies because they determined that their parent organization didn't have people with the required skills. This was an expensive effort that also created animosity with the parent company's employees. However, it did not result in a sustaining advantage and long-term success. Perhaps they did not have a critical mass of people for each strategic function. Or perhaps they did not organize the newly hired employees in a way that the right collaborations could emerge. Or the resistance from the parent employees had a dilutive effect on the efforts of the new employees and inhibited collaborators to emerge.

Observation 5: ***Dealers can remain important partners even as the business model changes.*** Dealers played a key role in the initial success of the S-Series and the customer experience it introduced. Less so for the i3. Saturn saw its new no-negotiation business model as the key component of the customer experience the company wanted to offer. They arrived at this model by being customer-centric and understanding that the dealer-customer price negotiation was creating significant friction in the customer relationship. But Saturn not being willing or able to take over the customer relationship and own it also recognized that the business model will create a problem in dealer economics. To achieve the desired outcome Saturn established a completely different dealer network and provided each member with a larger operating territory and thus less competition. This improved the dealer economics, thus alleviating the problem introduced by the company's new business model. The success of the S-Series made the dealers happy and led the dealers to increase their collaboration and support of the company.

The i3 had a more complex customer experience because it was a Battery Electric Vehicle that was going to be offered with a collection of services. It required the dealer to create a more complex value proposition for each prospective customer. Again, since BMW was not ready to bypass its dealers and own the relationship of the i-brand customers it needed to provide them with more education about the product and ways to gain a better understanding of every prospect. Neither of these was provided by BMW. Dealers tend to be transactional, so it was unlikely that they would invest themselves in obtaining this education and customer understanding. So it fell upon the customer to appreciate the virtues of the i3. The sales figures showed that this did not lead to stellar results.

Observation 6: ***To be considered successful, innovation must create value.*** Saturn and the i-brand innovated in automotive technology, business models, and customer experience—yet neither provided enduring valuc to the parent company. There are at least two ways to evaluate innovation. First, the applied innovations result in a successful standalone business. Second, the innovations benefit other businesses associated with the innovator. Identifying and quantifying value in multifaceted innovations that require extensive transformations is hard since it is not always clear what type of innovation or which of the transformations was responsible for the achieved success. Corporations must find ways to measure the value of the innovations they create and the transformations they undertake against the investments they make. Neither the Saturn Corporation nor the i-brand did. Moreover, their primary innovations (in Saturn's casc: plastic-paneled compact vehicles, workforce organizational structure, dealer network structure, no-negotiation pricing; in i-brand's case: vertically integrated vehicle design and manufacturing, vehicles and services designed around the integrated megacity mobility experience, vehicle-as-a-service) did not transfer back to the parent organizations to result in a broader corporate transformation.

7.3 Transforming to Deliver the Technology

Most incumbent OEMs have chosen to initially move from Category 1 to Category 2 by introducing Domain-Based Software-Defined Vehicles rather than to Category 4. They are transforming along the technology dimension and appear to be less concerned about being customer relationship owners. In general, because of their corporate cultures and their senior executives coming from engineering and manufacturing backgrounds, incumbent OEMs find technology transformations easier to undertake than business model transformations that impact the customer experience. They are electrifying their model lineup to address customer demand, respond to government mandates, or just compete with newcomer OEMs. Electrification provided the opportunity to rethink the vehicle's architecture, giving rise to the Software-Defined Vehicle. Having committed to developing Software-Defined Vehicles, a decision that places them in Category 2, incumbent OEMs must now determine whether to move to Category 4 or just remain in Category 2. Moving first from Category 1 to Category 2 or Category 3 and then to Category 4 will prove more expensive and painful than the decision to move directly to Category 4 because it will keep these OEMs in transformation mode over a longer period and the organizational uncertainty and friction that such mode entails.

The Software-Defined Vehicles they are developing, or have already introduced, use Domain-Based Architectures. Their efforts are at different maturity levels. They continue to separate vehicle development from customer experience beyond what is provided around the vehicle's infotainment system. Among other criteria, they must decide on the level of customer relationship they will try to attain. They have three choices: owning a large part of the customer relationship like Apple, Google, Amazon and a few other technology companies do today, owning small parts of the customer relationship like AT&T and Verizon do, which is where they are today, or owning little or no part of the customer relationship like Foxconn or Magna do (Simoudis, "Will Automakers Become Apple, AT&T, or Foxconn?," 2021). Many OEMs continue to outsource the development of infotainment systems for their Software-Defined Vehicles. Automakers like Renault and Volvo are adopting Google's Android Automotive operating system as the basis for the applications and services they will offer. Stellantis partnered with Amazon to build an in-vehicle user experience that can be customized for all its brands. Even with an updated third-party user experience, the customer experience of these automakers will not change. It appears that their customer relationship will remain with the dealer.

Remaining in Category 2 and not moving to Category 4 will mean that the OEM will only need to make changes in the technology organization. Though moving to Category 4 should be the goal, this will require broader organizational changes and changes in the OEM's value chain, and as said above, many OEMs prefer to first spend time in Category 2. To address that reality: the organizational transformation needed in Category 2 will depend on which part of the Zone-Based Software-Defined Vehicle's stack they decide to own to create the proper product differentiation. The organizational transformations they undertake and the people they bring into the organizations they form must enable them to achieve this goal. For everything else, they will need to find appropriate partners. We are already seeing several such partnerships emerging. For example, most automakers partner with public cloud providers such as AWS and Azure. Incumbents are setting a goal to develop 20 to 60 percent of the Software-Defined Vehicle's technology in-house and license the rest from their partners (*The Economist*, 2022). Depending on which parts of the Software-Defined Vehicle technology stack they want to own and the release timelines they announce, two organizational transformation approaches have emerged relating to the development of technologies associated with Zone-Based Software-Defined Vehicles.

The first approach is to form a group under the OEM's existing organizational structure whose mission is to create the Software-Defined Vehicles architecture and associated technology stack. This organization brings together under one roof the people with the areas of expertise to develop the selected components of the Zone-Based Software-Defined Vehicle stack and presumably establish the necessary agile processes and culture. BMW, Mercedes, and Hyundai are three OEMs that have taken this approach. According to Frank Weber, head of R&D at BMW, the company is following this approach for the development of its Neue Klasse platform and will introduce a new i3, along with several other models, starting in 2025 (Kacher, 2022). Hyundai's Software-Defined Vehicle program is developing its Integrated Modular Architecture and the ccOS software platform. This effort is headed by Chung Kook Park, the President of its R&D Division. Mercedes recruited Magnus Östberg, its chief software officer, from Aptiv to lead the Electric Software Hub organization that is developing the Mercedes Modular Architecture (MMA) and its successor platforms, as well as the MB.OS middleware (Paul, 2023; Ravi, 2022). The MMA will be installed in ICE vehicles to provide them with a new customer experience. After that, Mercedes will introduce electric vehicle-only platforms. These include the MB.EA, which will be installed in medium and large passenger cars, AMG.EA, a platform for performance electric vehicles, and VAN.EA for light commercial electric vehicles.

The second organizational transformation approach is to form a separate technology-focused company under which to hire people with the required areas of expertise. Toyota, VW, and Renault adopted this approach. Toyota formed Woven Planet in 2018. VW established CARIAD as a separate company in 2020. On November 2022 the Renault Group announced it will split into five different companies, with Ampere being the company responsible for the development and sale of the Software-Defined Vehicles (Torsoli, 2022). Few details have been provided about the reorganization to date other than that Google and Qualcomm will be Ampere's strategic partners in the development of the Software-Defined Vehicle models and the in-vehicle experience.

Interviews with these organizations confirmed the following four reasons for adopting this approach. First, it is easier to attract and hire employees from the high-technology industry under a new company with a lean structure than to hire them under the vast and hierarchical organizations of incumbent OEMs. The targeted candidates, many of whom have startup experience, prefer to work in companies with flatter organizational structures. Second, with an organization that is completely separated from the OEM, it is easier to create a work culture that comes closer to that of high-technology companies and institute the agile and lean processes necessary for the accelerated development of Software-Defined Vehicles that automakers need. Third, it is easier to be competitive in compensation issues. In addition to competing with newcomer OEMs for the same talent, incumbent OEMs must participate in a broad war for the right talent even as economies are slowing down. This is because the technologies that are key for the development of Software-Defined Vehicles, e.g., AI, cloud-computing, materials, etc., are equally important in several other industries. Compensation becomes an important weapon in this war for attracting and retaining the right employees. It is easier to create competitive compensation packages under a separate company than under the incumbent automaker. Depending on how the new company is set up, it may also be able to separately go public, providing an additional incentive for employees, but also for the parent corporation. Finally, it is easier to expand these companies in high-technology hubs under their standalone structure. For example, Woven Planet now has operations in Tokyo, Palo Alto, Seattle, and Ann Arbor. CARIAD has operations in Berlin, Ingolstadt, Seattle, and Shanghai. A problem that cannot be overlooked, however, is that the candidates being recruited from companies that are outside the automotive industry, while they may be experts in agile processes, are not familiar with the idiosyncrasies of automotive software, which can negatively impact the quality of the software they create.

As part of their organizational transformations associated with technology development, corporations are also acquiring startups. For example, in July 2021 Woven Planet acquired Lvl5, which was a division of Lyft and had a staff of over three hundred technical people (Doll, "Toyota's Woven Planet acquires Lyft's Level 5 self-driving division," 2021). Later that year it also acquired Renovo, a startup that was developing a middleware platform. In July 2022, Hyundai acquired 42dot, a startup that was developing an autonomous mobility platform (Kim, 2022). Unfortunately, many of these acquisitions do not succeed because of the culture clash that typically emerges between the startup's culture and that of the acquirer.

Targeting to move to Category 4 requires different organizational transformations. We point to the one taken by Ford and Renault. On March 2022 Ford announced its split into three different units that will operate under a single umbrella organization (Wayland, 2022). Under this transformation, Ford Blue will continue to focus on ICE vehicles, i.e., remain in Category 1; Ford P ro focuses on fleet vehicles and related services, i.e., is already in Category 4; and Model E focuses on Zone-Based Software-Defined Vehicles, i.e., it will become a member of the Category 4 segment that currently includes only newcomer OEMs. It is not yet clear how this will work in the long term even though in its most recent Capital Markets Day event Ford showed that it is already tracking each unit independently, implying that it is well on its way with this transformation (Simoudis, "Ford CMD: The Right Goals Are Set; But Can The Company Execute?," 2023). With the introduction of the F-150 Lightning and the Mach-E Domain-Based Software-Defined Vehicles that are sold through its existing dealer network and most of the other elements of the existing overall customer experience, Ford is also a member of Category 2. The first vehicle of the Model E unit will be introduced in 2025 and will have a new end-to-end customer experience whose details have not been revealed.

With its organizational transformation, Ford can achieve four goals. First, allow each unit to establish a culture that suits its goals. For example, Model E will try to establish a high-technology company culture to attract the needed talent. Jim Farley, Ford's CEO, believes that this is a key requirement for succeeding in the emerging electric vehicles market, implicitly admitting that Ford needs different talent to succeed in the Software-Defined Vehicle market (Boudette, 2022). The Model E unit is led by Doug Field, an executive Ford recruited from Tesla. Roz Ho, the unit's Chief Software Officer, was recruited from HP, and Gil Gur Arie, who leads analytics, was previously heading R&D for a unit of the Israeli Defense Forces. Second, because the three units will be under the same umbrella, they will be able to share certain technologies. The ADAS functionality called BlueCruise was cited as one such example. Third, the Model E unit will benefit from the cash flow and scale of Ford Blue to address its high capital requirements. Fourth, by having such a clear split between the three units, Ford puts itself in a better position to own the customers of Model E and Ford Pro.

Other incumbent OEMs that are moving to Category 4 include GM and Afeela, the Honda/Sony joint venture. To achieve its goal, GM has created technology development and customer strategy groups under its existing organizational structure. These groups bring together software, hardware, batteries, and customer experience, including the teams that develop the services that will be offered as part of the new customer experience. GM formed a new technology organization to develop its Ultium/Ultifi platforms. GM argued that within its vast engineering organization, it always had teams working on the technologies that are key ingredients to Software-Defined Vehicles. Over time it has grown these teams accordingly, but it didn't have to start from scratch. Many of the applications that today are developed for its Ultifi software platform have their roots in the OnStar offering or have been incubated within GM. The organization was initially led by a long-time GM executive, but more recently GM recruited Michael Abbott from Apple to oversee the effort (Colias, 2023). The decisions relating to the ownership of the customer relationship, including the services that will be offered under a new customer experience, are driven by GM's Digital Business group, which is headed by Alan Wexler, who is also leading strategy and innovation for the company, as well as Donald Chesnut, the company's Chief Customer Experience Officer. Wexler came to GM in 2019 from Publicis. Chesnut came to GM in 2021 from Mastercard.

In creating Afeela, Honda and Sony took a more radical organizational approach. They created a newcomer OEM out of incumbents in the automotive and consumer electronics industry incumbents. The company's top management consists of senior executives from both companies, as do many of the company's other employees. The company intends to own its customers by only offering long-term lease options for its vehicles.

7.4 Transforming to Create and Offer the Flagship Experience

The OEMs in Category 4 can monetize the customer through the Flagship Experience's customer journey and the Software-Defined Vehicle throughout the cradle-to-grave vehicle journey. Armed with these capacities, they need to determine where they will create value based on their understanding of each customer's needs. The OEMs in this category must not only offer value-adding services that are the result of deeply understanding each customer's needs and lead to customer monetization but must ensure that they become the preferred providers of these services. The SDV Services Marketplace must be populated with both *owned-and-operated services*, e.g., usage-based insurance, as well as services provided by partners, e.g., electric vehicle charging. For example, OnStar is one of GM's owned-and-operated services. Owned-and-operated services should capitalize on the OEM's key strengths, provide unique value, create barriers to entry for competitors, and enable the OEM to enhance its customer understanding while collecting valuable data. As Donald Chesnut, GM's Chief Experience Officer, stated to the author, "Not all income from the new vehicles has to come directly from the customer. Some could come from ecosystem partners."

Financial services companies will participate in the new value chains by offering frictionless payment technology to emerging driver commerce applications. Utility companies are already starting to partner with automakers (Johnson, August). Many of these industries, e.g., telco, energy, hospitality, digital services, and even retailing will be participating in mobility for the first time.

The OEMs in this category must create organizations that can accomplish this goal. To develop and offer such services, the automaker must bring together an organization that combines strategic decision-making, service ideation and development, partner recruitment, and customer understanding. This organization must form a strategic relationship with the OEM's finance organization because the business model change, starting with the OEM approaching their Parc as fleet and managing each Software-Defined Vehicle's cradle-to-grave vehicle journey, will have a critical impact on the company's financial position. In this area, we see three different approaches adopted by GM (Chesnut's Digital Business Group that scales the services that were incubated by Alan Wexler's organization; the same organization has also incubated and launched BrightDrop), Ford (Model E unit that is developing its services), and Afeela (newcomer OEM created as a joint venture by two incumbents).

The provided services must address the needs of each target customer segment. For example, could in-vehicle commerce result in a set of services that will appeal to a large set of customer segments (even if only involves purchasing coffee, small bites, as well as vehicle charging/gas and parking)? The newcomer Chinese OEMs Nio, BYD, and Xiaomi are already thinking in these terms. By adopting Google Automotive Services, incumbents such as Polestar, Volvo, Honda's Acura brand, and Renault decided that at least the in-cabin customer experience they will offer in their vehicles will be based on Google's vision.

Service ideation requires strong product management skills because it involves determining which services will be monetizable, identifying and designing the service's features that will compel the customer to see its value, and establishing the business model under which the service will be offered. It also requires a culture of co-innovation. As Stephan Durach, senior vice president of development at BMW's Connected Company, stated to the author, "The OEM will need the ability to quickly introduce in the vehicle features and services in response to post-sale market needs." The selected business models may involve direct monetization through transactions, subscriptions, or the collection and use of customer data by the automaker. Service development requires agile application implementation skills (described in Chapter 5). Determining which services to own and for which to recruit partners requires strategic thinking and strong business development capabilities. As was described in Chapter 6 under the FEAT methodology, customer understanding requires extensive and multifaceted data collection, large-scale analysis and insight utilization, and deep AI expertise. Establishing the new business models and incorporating them into the automaker's overall financial model will require significant analysis and negotiation between the OEM's various organizations. It is better accomplished among the OEM's most senior executives. Their adoption of such services will also require significant changes in the OEM's role across the value chain and could result in conflicts with existing and prospective partners that directly impact the automaker's ability to monetize the customer.

Consider GM's recent decision to eliminate Apple's CarPlay from its electric vehicles by 2024 and replace it with its in-cabin infotainment system and several of its owned-and-operated applications (George, "Everybody hates GM's decision to kill Apple CarPlay and Android Auto for its EVs," 2023). This is not the first time GM is attempting to equip its vehicles with an internally developed infotainment system as it tried to establish a stronger relationship with its customers. In 2011, it tried something similar with MyLink and the Cadillac User Experience (Hearst 2021; Lutz 2011). The Cadillac system was the most ambitious and was co-developed with Bosch. It was received poorly because it was unresponsive and poorly designed, leading to user frustration. Its problems were the result of poor technology choices, a lack of understanding of how mobile applications fit into the customer's digital life, and software implementation by automotive engineers rather than mobile application engineers. Remember that 2011 was the year Apple released the iPhone 4S running iOS 5, and Google released Android 4, which was first installed on the Samsung Galaxy Nexus. Their app stores were already established and a big success. Using the various proprietary and partner applications provided by Apple and Google, consumers were well on their way to moving their digital lives to increasingly capable smartphones. In 2014 Apple introduced CarPlay, which GM adopted in 2016. Android Auto was introduced in 2015, and GM adopted it in 2017. Today most GM vehicles are equipped with Android Automotive, an infotainment system developed by Google, and have a CarPlay option.

GM's strategic decision regarding CarPlay makes sense. Customer monetization requires customer understanding, i.e., customer-centricity, and customer understanding requires customer ownership so that the OEM can have control and ownership of the customer data. The vehicle's infotainment system is a rich data collector and conduit. For the customer to split their digital life experience between the smartphone and the automaker's infotainment system, GM must demonstrate daily that the in-cabin experience provided by its system is at least as good as what is offered by the smartphone. Tesla showed that this is possible. One may say that Tesla succeeded because its early customers never got used to CarPlay (or Android Auto), but it is through continuously updating its infotainment system that Tesla provides the equivalent services in a way that customers find compelling.

As they develop and roll out the Flagship Experience, OEMs must be prepared to compete fiercely with the large companies, most importantly Google, Apple, and Amazon, that understand customer journeys and already own and monetize consumers' digital life using a variety of business models and are also entering various aspects of the consumer's physical life, e.g., home, health, and more recently vehicle. For this reason, they must understand the lessons these companies have learned in the process of achieving and maintaining their leadership positions in an evolving digital landscape. Google monetized search, email, and maps, on both desktop and mobile, through advertising, transactions, and subscriptions. Several other applications followed. For Apple, it was iTunes and Safari on the iPhone, monetized primarily through transactions and subscriptions. In the process of offering these applications for free with advertising, both collected and owned as much customer data as possible. Customers continue to move the services because these companies provide ongoing value to their customers in exchange for at least additional data. Tesla followed in their footsteps in thinking that the customer's lifetime value extends beyond the vehicle price and the initial vehicle owner. Insurance, supercharging, connectivity, and ADAS services present additional revenue opportunities that extend beyond the initial vehicle owner. It is also leveraging the customer relationship for services relating to electrification and energy storage products.

7.5 Recommendations

As in the past, incumbent OEMs will face several challenges as a result of the transformations they already undertook to develop Software-Defined Vehicles and will face even more as they proceed to create and offer the Flagship Experience. Moving to Category 2 and then to Category 4 will lead to OEMs remaining in transformation mode for a longer period than moving directly to Category 4 and will prove more expensive than moving directly to Category 4. However, this is now unavoidable for incumbents since they have all committed to first move to Category 2. Potential exceptions could be considered Ford's Model E and Afeela, but these two should be viewed as newcomer OEMs rather than incumbents because they have been set up and positioned as completely independent organizations with no strong connection to their parent.

Clearly define the role of dealers. The decision to own the customer relationship and the need to control the vehicle during the vehicle journey to monetize both customer and vehicle on an ongoing basis will create conflict between the incumbent OEMs and their dealers. These decisions negatively impact the dealer's role in the established value chain and the dealer's economics. They will lose the economics of reselling used vehicles, and their services revenue will decrease because the Software-Defined Vehicles are more reliable, and their software can be updated over the air by the OEM. Today dealers have a thirty percent share of the vehicle service market (Kitman, 2022). The rest of the market is shared among chains such as AutoZone and Pep Boys, as well as independents. They will also lose the ability to use the variety of customer data they own across all the brands they represent. Today the dealers own the data relating to the inquiries and choices prospective customers make, the services they provide to acquired vehicles even when they are out of warranty, and data relating to parts purchases. In the future, the regions where OEMs choose to maintain their dealer networks but also sell and support their customers directly will have a coopetition relationship with dealers because they will be targeting the same customers with potentially the same products and services.

Mercedes is moving to a direct sales model in Europe (Posky, 2022). GM has already offered to buy out Cadillac dealers that are not willing to make the requisite investments and is planning to do the same with Buick dealerships. GM and Ford are requiring their dealers to invest before they can sell their Software-Defined

Vehicles (Trop, 2022). Investments will relate to the education of the sales and support staff, the installation of chargers and other equipment necessary for the support of Software-Defined Vehicles, and the adoption of software that will enable the improved collaboration between dealer and OEM to provide the end-to-end customer experience and make more efficient the supply chain. Dealers are not eager to make these investments, and some are already opting out (Martinez, 2023). Because the independent dealers are being bought by large corporations, e.g., Autonation, Lithia, etc., as they contemplate these model changes, the OEMs will have to negotiate with large corporations rather than small and midsize companies. This could make the negotiations difficult and require OEMs to make tough, and potentially expensive, decisions.

Even without owning the customer relationship or the vehicle and becoming the OEM's agent, the dealer has a role under the Flagship Experience. As Donald Chesnut, GM's Chief Experience Officer, stated to the author, "Dealers could provide OEMs with important requirements of what to include in the customer experience, particularly the post-sales experience. Some of these services will be offered through applications. But others will require the human touch. For example, OEMs will need to rely on dealers to educate customers, at least certain customer segments, on the customer experience offered by the Software-Defined Vehicle and ensure they can appreciate it."

With every innovation, the response from initial customers is critical. Enthusiasm won't guarantee that the overall target market will adopt the innovation, but it provides the first encouraging signs. For this reason, the organizations introducing and supporting an innovation pay particular attention to the early adopters. The early adopters of the Software-Defined Vehicles that will be introduced by incumbent OEMs will require a hands-on approach to ensure their understanding and appreciation of the vehicle, its role in accessing the Flagship Experience, and the ways this experience will benefit them. Dealers can play an important role in this effort. According to a recent global study by PwC, many consumer segments continue to prefer the in-store physical experience in the process of acquiring a vehicle (PwC, 2022). In a 2022 survey of 1,289 US-based people conducted by EVForward, 74 percent of those surveyed indicated that they would prefer to buy an EV from a dealer than directly from the automaker (Brock, 2022). Respondents said they value the in-person experience compared to the online one, the convenience of nearby showrooms the large dealer networks offer, and the ability to experience and buy a vehicle that is available on the dealer lot rather than waiting several weeks or months for the delivery of the ordered vehicle. The dealers' role regarding the servicing of Software-Defined Vehicles is less clear.

Quickly resolve conflicts in the value chain. The value chain that will be created around Software-Defined Vehicles and the Flagship Experience will include new partners, many of whom will be looking to monetize the customer. Google and Amazon, for example, are positioning themselves to play a role across the entire value chain of Software-Defined Vehicles from platforms to direct-to-customer interactions. Partnerships will be necessary for the OEM to offer a full Flagship Experience but will inevitably create conflicts. Conflicts will arise around data ownership, intellectual property ownership, relationship ownership, and unit economics. For example, as automakers focus on software and data to help them achieve higher margins, conflicts will likely arise with application developers, as we already see today in the smartphone world, and service providers, such as charging networks. Even if the OEM believes that in the beginning its long-term partners, such as its Tier 1 suppliers, may not present a value chain conflict, such partners are likely to be undergoing similar transformations that lead them to develop or acquire capabilities that will result in value chain conflicts. For example, Mobileye, which initially provided OEMs only with ADAS sensors for their Software-Defined Vehicles, now competes with OEMs around high-definition digital maps. This is why OEMs must clearly define which parts of the new value chain they need to dominate and where partnerships will benefit them in providing enduring value to their customers.

Consider, for example, the partnership between OEMs and car rental companies. After ordering electric vehicles from GM, Polestar, Tesla, and BYD, car rental companies Hertz and Sixt have started deploying these vehicles in the US and Europe. Hertz has ordered one hundred thousand Tesla Model 3 and Model Y, one hundred seventy-five thousand GM vehicles starting with the Bolt EV, and sixty-five thousand Polestar2 (Ferris, 2022). Sixt ordered one hundred thousand Atto3 software-defined electric vehicles from BYD (Mihalascu, 2022). Once properly trained, these companies' personnel will be able to help consumers familiarize themselves with the capabilities of these vehicles, thus minimizing the learning and psychological costs associated with the adoption of these vehicles. In this way, very similar to one of the roles dealers can play in the future, car rental company personnel play an important role in convincing potential prospective Software-Defined Vehicle customers to become adopters and believers. Consumers will be able to experience the features of these Software-Defined Vehicles over longer periods than the typical test drive. As part of this value exchange partnership, automakers receive relevant data from the use of these vehicles, including renter feedback, vehicle telemetry, and trip analytics. This can be considered analogous to the partnerships that connected home product companies have with hotel chains to test if consumers will adopt new products. OEMs can benefit greatly by understanding the psychology of new product adoption as they introduce the Flagship Experience with their clean sheet Software-Defined Vehicles (Gourville, 2006).

Prepare for a culture clash. As with the Saturn and the early days of the i-brand, the organizational changes necessary to bring to market the Software-Defined Vehicles with the Flagship Experience will create a clash of cultures. The differences between the established way of creating a vehicle today and the agile approach advocated for the clean sheet Software-Defined Vehicles are significant. The electromechanical vehicle is built by centralized, hierarchical, tightly controlled organizations whose groups are structured by area of expertise and working over timelines of four to seven years. Every vehicle detail is considered before production starts. Agile development assumes flat organizations of problem-solvers that are often decentralized and are becoming more so, particularly as software experts become digital nomads. Decentralized organizations are successful when there is trust among team members, as well as between management and employees. The members of these organizations expect that Continuous Integration/Continuous Delivery (CI/CD) is part of the organizational DNA and not just a slogan. The complexity and newness of Software-Defined Vehicles make CI/CD a necessity, but it must also be coupled with good automotive software expertise to avoid problems during the operation of Software-Defined Vehicles. In other words, it will be impossible to fully specify the features of the Software-Defined Vehicle and the services that will comprise the Flagship Experience.

The implications of agile development and CI/CD methodology result in the second cultural clash because they impact a vehicle's safety. The engineers and product managers who work in the smartphone industry are used to introducing new features and services that may have bugs. The smart device user expects to report and live with such bugs until the service's next release. However, buggy features and services can have safety implications for the user of the Software-Defined Vehicle. The automotive industry has established a regulated process for reporting and fixing vehicle problems. It involves recalls. According to the NHTSA, an OEM must correct an identified problem by repairing the vehicle at no cost to the customer, replacing the vehicle with an identical vehicle or similar model, or refunding the depreciated value of the vehicle.

The two cultures also differ in their urgency toward timelines and lifecycles. Incumbent OEMs built their operations on the aforementioned four- to seven-year lifecycle. A new complete model upgrade is introduced approximately every seven years with a refresh three to four years after the upgrade. The development of every vehicle component (software or hardware) works in lockstep and must adhere to the timeline imposed by this cycle. The Software-Defined Vehicle and associated Flagship Experience necessitate that the new organizations become comfortable with simultaneously operating under the three different lifecycles mentioned in Chapter 5: the software lifecycle, the sensor/consumer electronics and battery technology lifecycle, and the skateboard and body lifecycle. Existing employees who transfer into the organizations developing the Software-Defined Vehicles and the Flagship Experience will need to quickly adopt the practices surrounding the first two lifecycles because they imply a different way of working compared to what they are used to.

The third cultural clash will be the result of the customer-centric strategy the OEM will need to adopt. One cannot address a customer's ongoing needs without constant dedication to understanding them. Customer-centricity is not achieved by receiving feedback from a few focus groups. It implies constantly seeking feedback from the entire customer base, and even prospective customers, and acting on it. Therefore, for the Flagship Experience to be implemented and successfully deployed, OEM employees must be customer-driven. The data they collect, even when it involves complaints about a recently released product or service, help them improve the product itself and the process that drove its release. OEMs must bring their employees as close as possible to the "voice of the customer" and compel them to listen to it and reflect it on their products. This approach will be driven by the software engineers but also be accepted by the mechanical engineers who controlled the development of electromechanical vehicles and whose priority was to deliver the vehicle that was designed in a top-down manner with input from only a few focus groups.

The friction that results from the attempts to integrate two different cultures could lead to product delays, product quality issues, and even personnel departures. Moving first to Category 2 and then to Category 4 is not expected to lessen the degree of the culture clash compared to moving directly to Category 4. It is not clear that any one of the organizational structures being used for the development of the Software-Defined Vehicles will be more appropriate for minimizing the culture clash. Creating a new company like VW and Toyota did with CARIAD and Woven Planet, respectively, may reduce the degree of the clash at the employee level but could lead to a different level of scrutiny by the parent organization, as we are already seeing in these two organizations. Woven Planet announced a new structure that appears to bring it under closer control and supervision by Toyota. CARIAD changed CEOs twice in the last three years.

OEMs must continue to show commitment, provide ongoing support to the management teams that execute these transformations, and continue to invest to complete these transformations. These organizations should showcase the benefits of the new approaches and technologies even on conventional vehicles, as Ford is currently doing. Even though the organizations producing the ICE vehicles will remain the profit drivers in the short term and will thus enable the development of the Software-Defined Vehicles, the OEM's management should show unwavering support to the units tasked with the development of new vehicles and the new customer experience.

Hire the right people and the right number of people. Automakers are not short of software engineers. They have plenty of engineers working on the embedded systems that govern the microcontroller architectures. They are short of software engineers who understand cloud computing, microservices-based architectures, AI, and other relevant technologies that are at the core of modern software systems, and which are necessary for the implementation of Software-Defined Vehicles and the Flagship Experience. They are also short of executives who can lead the technology and business transformations and bring them to successful outcomes. The new employees must have a mobility mindset rather than a manufacturing mindset.

To fill their management ranks they have been hiring from outside the automotive industry. In addition to the hires mentioned earlier, in 2022 GM hired Jon Francis, who was previously with PayPal and Starbucks. In 2023 Stellantis created the Mobilisights business under Sanjiv Ghate, an executive who was hired from HERE Technologies, and VW hired Ricarda Heim, its Chief Data and Analytics Officer, from Microsoft. This new talent does not address another issue, though: that the responsibilities for these transformations are spread among too many executives. In particular, to make the Software-Defined Vehicle and Flagship Experience a reality under the current organizational structures many executives must work together: CEO, chief vehicle designer, chief software officer, chief data officer, chief information officer, chief customer officer, chief experience officer, chief strategy officer, and chief marketing officer. Not every OEM has executives fulfilling these roles. In some cases, an executive may perform more than one of these roles.

Software engineering will not be the only skillset OEMs will require for the implementation of the Software-Defined Vehicles and the Flagship Experience. For example, battery manufacturing requires a different skillset than radiator manufacturing. As a result, OEMs will need to reconcile the size of their existing manufacturing employee base with their future needs even if we are to assume that the annual production runs will remain the same. Many existing employees will have to be retrained since in many parts of the world OEMs will not be able to lay existing employees off because of employment law, union contracts, or both.

These programs require many people with these skills and preferably a high-technology industry background. Companies in the high-technology industry tend to hire good "athletes," i.e., their skills enable them to work in different roles depending on the problem to solve, rather than being unidimensional experts. The industry values risk-taking, innovative problem-solving, and informality. Tesla has hired many employees with these characteristics. Many of the company's innovations, starting with OTA updates, or in the way vehicles are manufactured, e.g., gigapresses, are possible because of the risk-taking that is encouraged. High-technology companies emphasize informality. This is not about the open-floor office layouts or the wearing of casual clothes. It has to do with the expected, or tolerated, behaviors. Their employees typically self-organize in teams, rather than waiting to be assigned by their immediate managers, a characteristic of the hierarchical organizations found in automakers. Decisions such as how a team works, during what hours it works, and how the team members communicate center on informality rather than procedures. Agility is not only practiced, but it is rewarded. It is demonstrated in the way products are designed, and redesigned, as market conditions change, new competitors emerge, and others pivot away. Teams may be quickly disbanded while new ones form to best respond to a situation.

OEMs are starting to recognize these traits and the number of people they must hire to create a critical mass within their organizations. Mercedes plans to hire over one thousand software engineers for its Software-Defined Vehicle program. CARIAD acquired several companies to access employees with the required skills. After naming Yves Bonnefont as its Chief Software Officer in 2021, Stellantis launched a "Data and Software Academy" intended to retrain more than one thousand of the firm's existing employees per year, as well as hiring talent worldwide, to have a team of four thousand five hundred software engineers by 2024. To compete effectively with newcomer OEMs and digital life leaders alike, incumbents must continue aggressively hiring executives and engineers that are at least as good as the employees of these competitors. They will need to find a way to attract such candidates, which also means that they will need to change their compensation structures and address the pushback they will likely receive from existing employees. In addition to competitive compensation, they will need to create an environment that presents attractive challenges and provides employees with a voice. This hiring practice should be independent of whether they want to remain in Category 2 or ultimately move to Category 4.

Changing the processes, technologies, people, and business model to deliver Software-Defined Vehicles and the Flagship Experience requires incumbent OEMs to undertake significant, multifaceted transformations. Many analysts believe that OEMs will not ultimately succeed in transforming and effectively competing with newcomers. Driven by the category where they want to end up, OEMs are choosing different routes to the transformations they select and how they perform them. OEMs that do not move beyond Category 2 may not face the dealer challenge, and the value chain challenge may not be as intense as it will be for OEMs transforming from Category 2 to Category 4. But as value moves from the vehicle's hardware to software, and the vehicle becomes configurable, enabling the customer's ongoing monetization during the post-sales period, it will be advantageous for OEMs to make Category 4 their goal. The outlined recommendations provide OEMs with an advanced warning system that makes them aware of the challenges and with ways to start addressing them.

CHAPTER 8

Conclusion

The automotive industry is undergoing its third transformation in its 135 year history. This transformation, spurred by the impacts of climate change, can help automakers adapt to a new mobility landscape and address other important goals. Software-Defined Vehicles are designed to take advantage of technological innovations to reduce or even eliminate transportation-related emissions and make Vision Zero an attainable reality while also enabling the OEM to create new revenue streams. Customers want vehicles that align with their values, address their needs, and fit in their lives. This leads automakers to a crossroads. Should they remain vehicle-centric during this radical transformation or use the transformation as also an opportunity to become customer-centric? The Flagship Experience is a comprehensive customer experience blueprint customer-centric OEMs can offer with their Software-Defined Vehicles to increase customer loyalty and capture a bigger piece of their customer lifetime value by identifying monetization opportunities throughout the customer relationship.

Even if they are to remain vehicle-centric, the customer requirements for the new vehicles are leading incumbent automakers into a classic innovator's dilemma. Should they develop such vehicles by using their existing architectures, maintaining the previously established supply chains, and employing their existing business models, or use the opportunity to develop clean sheet Software-Defined Vehicles and adopt new models? Choosing the former reduces their risk and capital requirements and enables them to address the demand earlier. But it limits the broader opportunity for the industry to change how it designs, manufactures, sells, and supports vehicles, as well as introduces new revenue streams. Choosing to develop Software-Defined Vehicles introduces a new set of dilemmas. OEMs must become customer-centric to determine what customer needs they must address and which of these needs will lead to monetization. They must also determine whether Domain-Based Architectures will be sufficient for achieving the monetization and customer loyalty goals they have established or whether more advanced architectures, e.g., Zone-Based, will be necessary. But as we showed in this book, vehicle technology alone will not suffice. A complete customer-centric rethinking of the customer experience must be coupled with the Software-Defined Vehicle. The Flagship Experience forces the automaker to consider both the customer journey through an end-to-end customer relationship and the vehicle journey through an equivalent cradle-to-grave vehicle relationship.

Over time, the incumbent OEMs choosing the lower-risk approach and remaining in Category 2 will be relegated to a role that is equivalent to today's Android handset manufacturers. Their only differentiation will potentially come from the vehicle design. Large scale manufacturing will not be a differentiator because vehicle contract manufacturers, e.g., Foxconn and Magna, are aggressively entering the market. These companies may, in turn, create their open vehicle designs, pair them with their technology stacks, and offer them to the market. But the decision to develop

Software-Defined Vehicles has led OEMs to commit large investments over a long period.

Many incumbent OEMs have announced major multiyear investments to develop Software-Defined Vehicles. While making these investments incumbents must deal with the cyclicality of the automotive market, geopolitical tensions, and increasing competition. The strength of the automotive market is correlated to the strength of the economy. Most recently we are seeing once again the impact of the rising interest rates and the slowdown in the world economy on the sales of privately owned vehicles, which are leading OEMs to offer more incentives and dealer inventories to rise. Geopolitical tensions, particularly those between the West and China, are impacting both vehicle sales and the automaker supply chains. Consider that VW derives 40 percent of its sales from China, whereas Mercedes and BMW derive 30 percent each. The competition from newcomer OEMs like Tesla, Rivian, and Nio, established automakers from China that are now entering the global market such as BYD, and MG is increasing. These challenges are not expected to disappear soon. Based on recent announcements, the decoupling of the supply chains that were built over the last thirty years between the West and China will accelerate. Today Chinese automakers are working aggressively to capture the European market's low and middle segments, and they plan to target the high-end segment as a second phase.

Becoming customer-centric will require OEMs to first understand customers in each of the vehicle's three contexts (i.e., as a transportation modality, as a workspace, and as an entertainment and socialization space) and address competitive challenges from technology companies. Consumers have invested in their digital life, particularly through their mobile devices. Technology companies like Google, Apple, and Amazon, which control major portions of digital life, want to extend this control to mobility, including within the vehicle. To this end, they want to control the vehicle's software stack and the monetization it enables. These technology companies are in the position to capture the lion's share of customer monetization in Software-Defined Vehicles by using the same playbook they are using to capture other parts of the consumer's physical life, e.g., home automation, while remaining the consumer's main digital life partners. With owned-and-operated services, such as auto insurance, that reflect their unique customer know-how and involve assets they fully control, OEMs will be able to incorporate themselves into their customers' digital lives and start their post-sales monetization.

Despite these challenges, OEMs must not alter their plans to develop and deploy Software-Defined Vehicles initially employing Domain-Based Architectures and, later on, Zone-Based. They must at least stick to the timelines they have announced, if not accelerate them. Their recent streak of profitable quarters and their ability to borrow enable them to continue making the investments they announced. Incumbents should continue selling ICE vehicles under their existing business models, work to maximize their margins from the sale of these vehicles, and use their balance sheet to fund their Software-Defined Vehicle efforts, as Ford is doing. They can also consider forming joint ventures with one or more partners, as Honda did with Sony to create Afeela.

Governments are offering OEMs support in the challenges they face. This assistance comes in the form of investments and often through regulation. The US and the EU have announced big investments that will impact Software-Defined Vehicles and the Flagship Experience. For example, the US CHIPS Act provides approximately $280 billion in funding to boost US research and manufacturing of semiconductors, including those used in clean sheet Software-Defined Vehicles. The US Inflation Reduction Act provides $400 billion toward clean technologies, including the technologies used in electric vehicles. These investments will have a broad impact. Software-Defined Vehicles at scale are not merely new technology or a new civil-engineering project like building the US interstate highway system. They impact many industries, social and political structures, national and global economics, security, political competition, and of course the global climate.

Data and AI are important areas that impact the Software-Defined Vehicle and the Flagship Experience and require government regulatory attention that will help both automakers and their customers. Today we are seeing different regulations state-by-state and country-by-country relating to both data and vehicles. China restricts what data can be captured by a vehicle's sensors to be used in maps and real-time navigation. The European Union and some US states, starting with California, focus more on consumer privacy. Regulatory inconsistency is problematic for the automaker because it impacts their investment decisions relating to scalability and confuses the customer, resulting in their hesitation to acquire the Software-Defined Vehicles. The discussion about how the government can regulate AI, and how such regulation will impact the OEMs' ability to understand customer needs and match them to recommended services, is just starting. Today it is driven primarily by generative AI and less so by the type of AI used by the intelligent applications in the Software-Defined Vehicle and the Flagship Experience.

The government's role in these areas is important because the data generated by Software-Defined Vehicles, mobility infrastructures, mobility-related digital platforms, and customers themselves can also be applied to goals outside the vehicle. Data and AI-driven insights can be used to make urban mobility more convenient and cost-effective, improve traffic safety, and increase the use and monetization of transportation infrastructure. However, because this data can be used to compromise privacy and even endanger personal safety, governments must respect the cultural sensitivities of their citizens and work with OEMs and their ecosystems to properly address any misgivings. Surveys have revealed that Americans are particularly distrustful of government initiatives, including those involving the gathering and use of any type of data, while Japanese, Korean, and Chinese are more trusting. Europeans are vigilant about data that impact their privacy. Governments must strive to understand how data is used and will be used by the AI applications that are part of the Software-Defined Vehicle and the Flagship Experience. In this way, governments will be able to offer their input and make recommendations on how to avoid potential problems before they arise or before they are blown out of proportion by individuals or organizations that promote specific agendas. Since data is important to the AI applications that help OEMs understand customer needs, they must always be transparent with customers on how the data is used. Without such an understanding of how the data that is generated by the Software-Defined Vehicles is used, whether and how it is shared, and with what other data it is combined, the average customer will always be concerned and likely predisposed to restrict its acquisition and use in any part of the Flagship Experience.

Regardless of how they benefit from government assistance, OEMs need to find new opportunities to monetize the Software-Defined Vehicles they will sell. Results by Tesla and relevant projections by GM and Ford show that the annual post-sale services revenue on average can be equivalent to two or three additional monthly vehicle payments. More importantly, the profit margin on this revenue is higher than the margin from vehicles. With proper design, the Flagship Experience can help OEMs achieve such revenue goals regardless of the vehicle class. To achieve these goals they must establish a strong relationship with each customer that will lead to enduring customer loyalty, become customer-centric to understand the customer's mobility-related needs beyond just selling them a vehicle, and create value propositions that can be monetized during the length of the relationship.

Technology development is an important hurdle OEMs will have to overcome. Zone-Based and Central Computer Architectures, because of the computing power they incorporate, their data generation capability, and the level of component control they can impose, are the best fit for the Flagship Experience, and we are already seeing early cases of what happens if they don't get the technology right. In VW's case, the software stack developed by their CARIAD organization lacked features, had quality problems, and could not properly execute OTA updates, resulting in poor customer perception and low satisfaction. Renault and Stellantis adopted Google's and Amazon's technology stacks respectively, and Renault will also adopt Google's applications. By doing so they are ceding significant control of the customer experience.

Automakers that plan to collaborate with partners in the development of these technology stacks, particularly software platforms and AI applications, will need to evaluate what they will be giving up in terms of technology know-how and new revenue streams and what they will be gaining other than expediting their time to market and potentially decreasing their upfront development costs. Premium automakers developing Software-Defined Vehicles (e.g., Mercedes) volume automakers (e.g., GM and VW) and smaller automakers (e.g., Honda) will approach these issues differently.

Several versions of platforms used by Software-Defined Vehicle models and the Flagship Experience associated with each model will need to be developed, tested, and modified, with each iteration being capital-intensive. Existing partners will need to be convinced to transform and make the necessary investments to participate in the value chains proposed by the automakers. New partners will need to be recruited. What will be the role of Tier 1 suppliers that have been their traditional partners as new partners are recruited? Companies such as Bosch, Continental, and ZF are starting to create new mobility businesses. Bosch offers micromobility services to consumers, while ZF is offering autonomous vehicles. Operational agility, the ability to experiment continuously and quickly assimilate the results of these experiments, and investing prudently will become core characteristics of successful OEMs. Most OEMs have invested in technologies relating to online sales, and most recently in next-generation vehicle cockpits.

Addressing transportation's impact on climate change, increasing customer loyalty, and capturing a larger percentage of the customer lifetime value than today are the three main drivers of the industry's radical transformation. To better understand the transformations automakers need to undertake to achieve their goal, we organized OEMs into four categories. Most incumbent automakers have chosen to first transform technologically by developing Software-Defined Vehicles, which has in turn affected their options and delayed their timelines for organizational and business model transformations. GM and Mercedes are actively considering how to offer monetizable services under the conventional automotive customer experience that involves their dealer networks. Ford's Model E unit and Afeela have undertaken the development of Software-Defined Vehicles at the same time as reinventing the customer experience, which has enabled them to establish direct-to-consumer relations with associated monetization paths.

Many believe that the incumbent automakers will not be able to successfully compete against the newcomers, including the Chinese companies, regarding vehicle technology and customer experience. Understanding what went wrong with past transformation efforts combined with the provided recommendations should help OEMs. The book has argued that developing Software-Defined Vehicles, particularly vehicles with Zone-Based or Central Computer Architectures, together with the Flagship Experience will enable incumbent OEMs to offer vehicles that can be personalized easily and offered at competitive prices, in the $25,000 to $35,000 range. This will be a particularly important characteristic at a time when low-priced Chinese retrofit and clean sheet Software-Defined Vehicles are entering the European market, and despite regulatory restrictions today, they will likely enter the US market as well soon.

The Zone-Based and Central Computer Architectures will also make possible new manufacturing technologies that will positively impact the automakers' profit margins. Because of the capabilities provided by their advanced architectures, it will be possible to introduce these vehicles with fewer features. In addition to contributing to the introductory price, this approach will simplify the OEM's model lineup. Additional features can be offered through the Flagship Experience as the customer's mobility behavior and needs are understood, monetized, preferably using subscription models, and installed via OTA software updates. Between initial price and personalization ability, these vehicles should appeal to larger customer segments rather than only the high-end affluent segment, as is the case today with most electric vehicles.

Producing competitive Software-Defined Vehicles and offering a Flagship Experience that instills customer loyalty and leads to profitable and ongoing monetization requires automakers to adopt an all-in approach, create agile organizations, and change their business models. Hiring the right people to execute these organizational, business model, and technology transformations take time and capital. The candidates must see and appreciate the automaker's commitment. None of these challenges will be easy to overcome, particularly for an industry that hasn't changed in several decades, with past attempts to transform it meeting limited success, that is capital intensive with relatively low-profit margins, lacks the people with the necessary expertise, and faces competition both from agile newcomers and technology companies with strong balance sheets. The transformations described in the book will take time, and not every OEM will be able to break from the existing model as an integrator of systems produced by its Tier 1 partners.

The Flagship Experience will require OEMs to think differently. They will need to become customer-centric, a move that will impact them financially and has the potential to dramatically change their relationships with their dealers and other channels. Customer centricity involves collecting customer data, but it is not only about that. Customer centricity is about understanding the customer needs and acting, ideally proactively, to address them. Starting with each journey element, automakers will need to customize journeys at the individual vehicle level utilizing the data they collect and the AI analyses they perform. They will need to understand how to augment customer journeys continuously and dynamically with the lessons they derive from their customer interactions. Creating a customer journey that addresses the end-to-end customer relationship and coupling these journeys with cradle-to-grave vehicle journeys will require the OEM to rethink and even reconfigure the entire ecosystem from suppliers to post-sales partners. It will require the OEM to determine whether to understand each vehicle user's needs and preferences in detail or try to operate using generic customer profiles. They will need to also determine whether and how they will share a customer relationship with other members of their ecosystem, and, if so, which components. Finally, they will need to assess how these changes will impact them financially in the short- and long-term.

Properly leveraging the customer understanding they achieve to successfully implement the Flagship Experience will require OEMs to utilize the FEAT framework. It will enable them to create new models for defining, dynamically extending, monetizing the post-sales experience, and developing the right value propositions for each customer. FEAT enables the OEM to understand each customer's needs and lifetime value, and incorporating the right value propositions in the customer journey has the potential to address the customer needs with the automaker's financial objectives. If they don't get the value propositions right, then they risk losing their customers' loyalty in addition to monetization. Without attaining a clear-cut level of Software-Defined Electric Vehicle purchase risk acceptability, it will be nearly impossible to get the 60 percent of US households with two or more ICE vehicles to feel confident enough to become EV-only families. Even among current EV owners, only about 10 percent are EV-only households.

The success of the Flagship Experience will not be a race for which car has the most features, but which vehicle can incorporate the features each of its owners' (initial and subsequent ones) needs and wants. Its monetization will depend on whether the value provided by each service incorporated into the Flagship Experience is a service that the customer justifies paying for.

Will a better and more personalized customer experience be appreciated by customers and enable OEMs to better monetize them after the vehicle is acquired? If so, will every customer segment show the same appreciation? Will the personalizable customer experience that is enabled by the Software-Defined Vehicle influence their faster adoption or is electrification alone going to be the feature customers focus on the most? Are there drawbacks to OEMs attempting to monetize the customer during the vehicle's entire lifecycle rather than at the time the vehicle is acquired?

Regardless of what motivates the automotive industry's transformation, be it climate change, technology innovations, competition from newcomer automakers and mobility services companies, or changing customer behaviors, the result will be a very different industry from the one we have come to know over the past several decades. Automakers find themselves at a crossroads. Out of this transformative period, new leaders will emerge, and current leaders may lose their position or even be consolidated.

Every member of the automotive value chain and more generally of the mobility value chain ought to consider their role in new mobility. This implies considering their role in the value chains that are already emerging and others that have the potential to emerge, the investments they want to make the and intellectual property they want and need to own to ensure their future success, and more importantly, the way they approach the new mobility customer.

The Software-Defined Vehicle provides automakers with the opportunity to address the needs of their industry's radical transformation. The Flagship Experience allows them to move on from the crossroads they are facing and become customer-centric. Through the Flagship Experience, they can create end-to-end relationships with customers and vehicles. The Flagship Experience's blueprint prescribes what this customer experience can be, what it will take to implement it, and how it will be monetized across many vehicles and owners. Each automaker, even each brand, will need to decide how to interpret it. But they first need to decide what road at the crossroads to select. There has never been a better time to make this decision.

References

Chapter 1

Capparella, J. (2022, April 5). *Honda and GM Deepen Ties, Promise 'New Series' of Affordable EVs.* Retrieved from Car and Driver: https://www.caranddriver.com/news/a39636625/honda-gm-affordable-ev-plans/.

Ewing, J. (2023, June 26). Retrieved from *The New York Times*: https://www.nytimes.com/2023/06/27/business/energy-environment/tesla-gm-ford-charging-electric-vehicles.html.

Lienert, P. (2022, October 25). *Automakers to Double Spending on EVs, Batteries to $1.2 Trillion by 2030.* Retrieved from Reuters: https://www.reuters.com/technology/exclusive-automakers-double-spending-evs-batteries-12-trillion-by-2030-2022-10-21/.

Muller, J. (2023, January 20). *Hertz to help cities go electric, starting in Denver.* Retrieved from Axios: https://www.axios.com/2023/01/20/hertz-electric-vehicles-cars-denver.

Ohno, T., & Bodek, N. (2021). *Toyota Production System: Beyond Large-Scale Production.* Taylor and Francis.

Chapter 2

Arriva. (2021). *A new era for public transport–adapting to new passenger behaviours and expectations.* London: Arriva.

Bennett, S. (2023, June 8). *Last Mile Delivery Statistics 2023.* Retrieved from https://webinarcare.com/best-last-mile-delivery-software/last-mile-delivery-statistics/.

Berg, N. (2022, January 26). *This map shows the dozens of U.S. cities that will get new public transit in 2022.* Retrieved from Fast Company: https://www.fastcompany.com/90715924/this-map-shows-the-dozens-of-u-s-cities-that-will-get-new-public-transit-in-2022.

Bowman, C. P. (2022, November 17). *Coronavirus Moving Study: People Left Big Cities, Temporary Moves Spiked In First 6 Months of COVID-19 Pandemic.* Retrieved from https://www.mymove.com/moving/coronavirus-moving-trends/.

Boyd, B. (2021). *Urbanization and the Mass Movement of People to Cities.* Retrieved from Grayline: https://graylinegroup.com/urbanization-catalyst-overview/.

Brewster, M. (2022, April 27). *Annual Retail Trade Survey Shows Impact of Online Shopping on Retail Sales During COVID-19 Pandemic.* Retrieved from United States Census Bureau: https://www.census.gov/library/stories/2022/04/ecommerce-sales-surged-during-pandemic.html.

Cantor, C., Soulopoulos, N., Fisher, R., O'Donovan, A., and Cheung, A. (2022, September 21). *Bloomberg NEF.* Retrieved from Zero-Emission Vehicles Progress Dashboard: https://assets.bbhub.io/professional/sites/24/BloombergNEF-ZEV-Dashboard-Sep-2022.pdf.

Casselman, B. (2021, September 25). *Pandemic Wave of Automation May Be Bad News for Workers.* Retrieved from The New York Times: https://www.nytimes.com/2021/07/03/business/economy/automation-workers-robots-pandemic.html?

Dizikes, P. (2023, April 28). *Study: Covid-19 has reduced diverse urban interactions.* Retrieved from MIT News: https://news.mit.edu/2023/study-covid-19-reduced-diverse-urban-interactions-0428.

Duffy, R. (2020, May 22). *Nuro and TuSimple Continue Service During Coronavirus.* Retrieved from https://www.emergingtechbrew.com/stories/2020/03/27/nuro-tusimple-continue-service-coronavirus.

Frey, W. H. (2021). *Pandemic population change across metro America: Accelerated migration, less immigration, fewer births and more deaths.* Washington, DC: Brookings.

Grant, A. (2022, November 29). *This $220 Billion Market Opens Up a Path for Driverless Cars.* Retrieved from Bloomberg: https://www.bloomberg.com/news/newsletters/2022-11-29/this-220-billion-market-opens-up-a-path-for-driverless-cars.

Irle, R. (2022). *Global EV Sales for 2022 H1.* Retrieved from EV Volumes: https://www.ev-volumes.com/news/global-ev-sales-for-2022-h1/.

Kober, E. (2022, January 20). *How Large Cities Can Grow Denser and Flourish: What the 2020 Census Reveals About Urban Sprawl.* Retrieved from Manhattan Institute: https://manhattan.institute/article/how-large-cities-can-grow-denser-and-flourish-what-the-2020-census-reveals-about-urban-sprawl.

Lufkin, B. (2021, June 15). *Is the great digital-nomad workforce actually coming?.* Retrieved from BBC: https://www.bbc.com/worklife/article/20210615-is-the-great-digital-nomad-workforce-actually-coming.

North American Bikeshare and Scootershare Association. (2022). *3rd Annual Shared Micromobility State of the Industry Report.* Long Beach: North American Bikeshare and Scootershare Association.

Peters, A. (2020, January 30). *Here are 11 more cities that have joined the car-free revolution.* Retrieved from Fast Company : fastcompany.com/90456075/here-are-11-more-neighborhoods-that-have-joined-the-car-free-revolution.

Simoudis, E. (2020). *Transportation Transformation.* Menlo Park: Corporate Innovators.

Smith, D. (2020, July 7). *Urban to Suburban: The Growing Shirt to the Suburbs as Covid-19 Changes the Way People Live.* Retrieved from Chushman & Wakefield: https://www.cushmanwakefield.com/en/insights/the-edge/urban-to-suburban-the-growing-shift-to-the-suburbs-as-covid-19-changes-the-way-people-live?

Smith, H. (2021, October 31). *Does Future Miami Really Flood Like That?* Retrieved from Sierra: https://www.sierraclub.org/sierra/does-future-miami-really-flood-reminiscence-fact-check-movie-science.

Stewart, D., Cotret, Y. R., and Davidson, I. (2021). *Post-pandemic traffic jams: What they might mean for tech, media, and telecom.* Retrieved from Deloitte: https://www2.deloitte.com/ca/en/pages/technology-media-and-telecommunications/articles/post-pandemic-traffic-jams.html.

The Economist. (2021, January 28). *How the pandemic reversed old migration patterns in Europe.* Retrieved from The Economist: https://www.economist.com/europe/2021/01/28/how-the-pandemic-reversed-old-migration-patterns-in-europe.

Wilson, K. (2022, September 27). *Study: Downtown Congestion is Still Down—And It Could Help Explain Roadway Dangers.* Retrieved from StreetsBlog: https://usa.streetsblog.org/2022/09/27/study-downtown-congestion-is-still-down-and-it-could-help-explain-roadway-dangers

Yu, Y., & Melgar, L. (2021, May 11). Americans Up and Moved During the Pandemic. Retrieved from *The Wall Street Journal*: https://www.wsj.com/articles/americans-up-and-moved-during-the-pandemic-heres-where-they-went-11620734566.

Chapter 3

Davis, K. (2022). *Margin is King in EVs.* London: IHS Markit.

Fader, P. (2012). *Customer Centricity.* Philadelphia: Wharton School Press.

Guillaume, G. (2022, April 27). *Faurecia-Aptoide Signs up Mercedes in Infotainment Win Over Google.* Retrieved from Reuters: https://www.reuters.com/technology/faurecia-aptoide-signs-up-mercedes-infotainment-win-over-google-2022-04-28/.

Huang, J., & O'Toole, T. (2020). *Customer Loyalty: The New Generation.* New York: McKinsey & Co.

Hughes, J., Chapnick, D., Block, I., & Ray, S. (2021, September 26). *What is Customer-Centricity, and Why Does It Matter?* Retrieved from California Management Review: https://cmr.berkeley.edu/2021/09/what-is-customer-centricity-and-why-does-it-matter/.

Huntin, B. (2022, October 9). *The Battle for Control of the Dashboard.* Retrieved from Car and Driver: https://www.caranddriver.com/features/a41521644/the-automotive-infotainment-battle/.

Lambert, F. (2023, July 4). *Tesla Wants to Pay You to Driver its Cars This Summer.* Retrieved from Electrek: https://electrek.co/2023/07/04/tesla-wants-to-pay-you-to-drive-its-cars-this-summer/.

LaReau, J. L. (2022, July 16). *General Motors offers rebate on 2023 Cadillac Lyriq if drivers sign NDA, agree to tracking.* Retrieved from Detroit Free Press: https://www.usatoday.com/story/money/cars/2022/07/16/gm-offers-rebate-cadillac-lyriq-drivers-tracking/10076785002/.

Lee, T. B. (2018, October 10). *Tesla remotely extends car batteries to help with Hurricane Michael.* Retrieved from Ars Technica: https://arstechnica.com/cars/2018/10/tesla-remotely-extends-car-batteries-to-help-with-hurricane-michael/.

Libby, T. (2022, August 1). *May 2022 Tesla brand loyalty more than doubles year-over-year and leads all brands industry-wide.* Retrieved from S&P Global Mobility: https://www.spglobal.com/mobility/en/research-analysis/may-2022-tesla-brand-loyalty-more-than-doubles-yearoveryear.html.

Michalakis, N. (2021, December 2). *Programmable Mobility.* Retrieved from TechCrunch Tokyo: https://www.youtube.com/watch?v=t_QXulR_6aE.

Mims, C. (2022, October 1). *The Next Big Battle Between Google and Apple Is for the Soul of Your Car.* Retrieved from The Wall Street Journal: https://www.wsj.com/articles/the-next-big-battle-between-google-and-apple-is-for-the-soul-of-your-car-11664596817.

Nellis, S., and White, J. (2022, June 7). *Analysis: Apple's Next Frontier Is Your Car's Dashboard.* Retrieved from Reuters: https://www.reuters.com/technology/apples-next-frontier-is-your-cars-dashboard-2022-06-07/.

PwC. (2020). *Digital Auto Report 2020.* Munich: PwC Strategy&.

Schrader, S. (2022, September 21). *2024 Ford Mustang Digital Gauges Have an '80s Fox Body Cluster Mode.* Retrieved from The Drive: https://www.thedrive.com/news/2024-ford-mustang-digital-gauges-can-mimic-80s-fox-body-cluster.

Shah, V. (2022, October 8). *Android Automotive: Who's partnering with Google?* Retrieved from Car Expert: https://www.carexpert.com.au/car-news/android-automotive-whos-partnering-with-google.

Stumppf, R. (2021, December 11). *Toyota Made Its Key Fob Remote Start Into a Subscription Service.* Retrieved from https://www.thedrive.com/news/43329/toyota-made-its-key-fob-remote-start-into-a-subscription-service: The Drive.

Weintraub, S. (2022, January 26). *Ford Pro pairs new E-Transit/F-150 Lightning with VIIZR Salesforce telematics over wine in Sonoma.* Retrieved from Electrek: https://electrek.co/2022/01/26/ford-pro-pairs-new-e-transit-f-150-lightning-with-viizr-salesforce-telematics-over-wine-in-sonoma/.

Chapter 4

Bellon, T., Jin, H., & Shepardson, D. (2022, February 18). *Tesla software updates allow quick fixes - and taking risks.* Retrieved from Reuters: https://www.reuters.com/business/autos-transportation/tesla-software-updates-allow-quick-fixes-taking-risks-2022-02-18/.

Capelle, T. (2023, March 30). *Testing GTP3.5 vs GPT4: Which Model Writes Better Code?* Retrieved from Weights & Biases: https://wandb.ai/capecape/gpt3vsgpt4/reports/Testing-GTP3-5-vs-GPT4-Which-Model-Writes-Better-Code---VmlldzozODAzMzQz.

Eloundou, T., Manning, S., Mishkin, P., and Rock, D. (2023, March 23). *GPTs are GPTs: An Early Look at the Labor Market Impact Potential of Large Language Models.* Retrieved from Arxiv Econ: https://arxiv.org/abs/2303.10130.

Foote, B. (2021, July 29). *Ford CEO Jim Farley Says Build to Order Paradigm is the Company's Future.* Retrieved from Ford Authority: https://fordauthority.com/2021/07/ford-ceo-jim-farley-says-build-to-order-paradigm-is-companys-future/.

Fox, E. (2022, February 4). *Tesla Publishes Patent for Global Headlamps Customizable to All Countries' Requirements.* Retrieved from Tesmanian: https://www.tesmanian.com/blogs/tesmanian-blog/tesla-has-published-a-patent-for-global-headlamps-to-make-them-customizable-to-all-countries-requirements?TrucksFoT.

George, P. (2022, December 14). *Volkswagen has given a name to its pain, and it is 'software'.* Retrieved from The Verge: https://www.theverge.com/2022/12/14/23508088/volkswagen-software-id4-bug-problem-smartphone.

George, P. (2023, January 23). *Sony and Honda's EV goes where the Apple Car never did.* Retrieved from The Verge: https://www.theverge.com/2023/1/23/23564431/sony-honda-ev-afeela-apple-subscription-lease-software.

IHS Markit. (2020). *Automotive OTA Deployment Strategies.* Detroit: IHS Markit.

Johnson, J. (2022, July 31). *Tesla's Multi Ton Giga Press - Tesla's Ultimate Advantage.* Retrieved from Torque News: https://www.torquenews.com/14335/teslas-multi-ton-giga-press-teslas-ultimate-advantage.

Jost, K. (2021, March 15). *Nio Expert Discusses the Company's FOTA Learnings.* Retrieved from Futurride: https://futurride.

com/2021/03/15/nio-expert-discusses-the-companys-fota-learnings/.

Kane, M. (2021, May 7). *Elecrical Architecture Comparison.* Retrieved from InsideEVs: https://insideevs.com/news/506048/tesla-ford-vw-electrical-architectures/.

Koster, A., Arora, A., & Quinn, M. (2021). *Chasing the Software-Defined Dream Car.* Boston: Boston Consulting Group.

Lambert, F. (2020, December 30). *Tesla achieved supply optimization at Gigafactory Shanghai, now says cost is linked to materials.* Retrieved from Electrek: https://electrek.co/2020/12/30/tesla-achieved-supply-optimization-gigafactory-shanghai-cost-materials/.

Lambert, F. (2021, February 4). *Tesla releases video of Giga Press in action producing giant single-piece rear body.* Retrieved from Electrek: https://electrek.co/2021/02/04/tesla-video-giga-press-in-action-producing-giant-single-piece-rear-body/.

Lienert, P., & Bellon, T. (2021, November 10). *Global Carmakers Now Target $515 Billion for EVs, Batteries.* Retrieved from Reuters: https://www.reuters.com/business/autos-transportation/exclusive-global-carmakers-now-target-515-billion-evs-batteries-2021-11-10/.

Miller, J. (2020, August 30). *Electric car costs to remain higher than traditional engines.* Retrieved from Financial Times: https://www.ft.com/content/a7e58ce7-4fab-424a-b1fa-f833ce948cb7.

Nellis, S. (2022, January 13). *General Motors taps three Qualcomm chips to power its Ultra Cruise feature.* Retrieved from Reuters: https://www.reuters.com/markets/deals/general-motors-taps-three-qualcomm-chips-power-its-ultra-cruise-feature-2022-01-06/.

Novet, J. (2019, March 16). *In picking Microsoft's cloud, Volkswagen shows that even carmakers have some fear of Amazon.* Retrieved from CNBC: https://www.cnbc.com/2019/03/16/why-volkswagen-chose-microsoft-azure-over-aws.html.

Partridge, J. (2021, May 9). *This article is more than 2 years old Electric cars 'will be cheaper to produce than fossil fuel vehicles by 2027'.* Retrieved from The Guardian: https://www.theguardian.com/business/2021/may/09/electric-cars-will-be-cheaper-to-produce-than-fossil-fuel-vehicles-by-2027.

Penthin, S., and Landgrebe, C. (2019). *Software-over-the-Air (SOTA) – Saving costs and improving customer experience.* Frankfurt: Software-over-the-Air (SOTA) – Saving costs and improving customer experience.

Tao, M. (2022, January 21). *Pony.ai unveils new autonomous computing unit built on Nvidia Drive Orin .* Retrieved from Robotics and Automation News: https://roboticsandautomationnews.com/2022/01/21/pony-ai-unveils-new-autonomous-computing-unit-built-on-nvidia-drive-orin/48502/#:~:text=Nvidia%20Drive%20Orin%20achieves%20254,accelerator%20(NVDLA)%20toolchain%20support.

Warren, T. (2023, March 22). *GitHub Copilot gets a new ChatGPT-like assistant to help developers write and fix code.* Retrieved from The Verge: https://www.theverge.com/2023/3/22/23651456/github-copilot-x-gpt-4-code-chat-voice-support.

Wheatley, M. (2019, April 23). *Ford partners with Amazon to build cloud services for connected cars.* Retrieved from Silicon Angle: https://siliconangle.com/2019/04/23/ford-partners-amazon-build-cloud-services-connected-cars/.

Wiggers, K. (2021, April 12). *Nvidia debuts Drive Atlan system-on-chip for autonomous vehicles.* Retrieved from Venturebeat: https://venturebeat.com/business/nvidia-debuts-drive-atlan-system-on-chip-for-autonomous-vehicles/.

WikiChip. (2019, April 22). *FSD Chip-Tesla.* Retrieved from WikiChip: https://en.wikichip.org/wiki/tesla_(car_company)/fsd_chip.

Wikipedia. (n.d.). *Low-Voltage Differential Signaling.* Retrieved from Wikipedia: https://www.wikiwand.com/en/Low-voltage_differential_signaling.

Chapter 5

Butters, J. (2023, May 14). *Auto industry software cure-all may cause headaches.* Retrieved from Autoomotive News: autonews.com/commentary/beware-software-auto-industry-cure-all-may-cause-headaches.

Eclipse Foundation. (2022, March 8). *The Eclipse Foundation Launches New Working Group for Software-Defined Vehicles.* Retrieved from Eclipse Foundation: https://newsroom.eclipse.org/news/announcements/eclipse-foundation-launches-new-working-group-software-defined-vehicles.

George, P. (2022, December 14). *Volkswagen has given a name to its pain, and it is 'software'.* Retrieved from The Verge: https://www.theverge.com/2022/12/14/23508088/volkswagen-software-id4-bug-problem-smartphone.

Rasmussen, B. (2023, May 11). *Volvo Cars delays EX90 production start.* Retrieved from Reuters: https://www.reuters.com/business/autos-transportation/volvo-cars-delays-production-volvo-ex90-2023-05-11/?.

Singh, M. (2021, August 20). *This is what Tesla uses for Software Development.* Retrieved from Medium: https://preettheman.medium.com/this-is-what-tesla-uses-for-software-development-190e85ef2722.

Waldersee, V. (2023, May 11). *Volkswagen CFO: Next-generation software platform to come towards 2027, 2028.* Retrieved from Reuters: https://www.reuters.com/technology/volkswagen-cfo-next-generation-software-platform-come-towards-2027-2028-2023-05-11/?.

Chapter 6

Aftermarket News. (2022, January 6). *Visteon Unveils Automotive App Store for the Connected Car.* Retrieved from Aftermarket News: https://www.aftermarketnews.com/visteon-unveils-automotive-appstore-for-the-connected-car/.

Burnham, A., Gohlke, D., Rush, L., et. al.(2021). *Comprehensive Total Cost of Ownership Quantification for Vehicles with Different Size Classes and Powertrains.* Argonne: Department of Energy.

Capparella, J. (2022, April 5). *Honda and GM Deepen Ties, Promise 'New Series' of Affordable EVs.* Retrieved from Car and Driver: https://www.caranddriver.com/news/a39636625/honda-gm-affordable-ev-plans/.

Cushing, C. (2023, April 17). *China's Li Auto aims to nearly triple electric model line-up to 11 by 2025.* Retrieved from Reuters: https://www.reuters.com/business/autos-transportation/li-auto-wants-build-over-3000-charging-stations-across-china-by-2025-2023-04-18/.

Doll, S. (2021, October 13). *How much is a Tesla lease?* Retrieved from Electrek: https://electrek.co/2021/10/13/tesla-lease-deals-a-complete-guide/

George, P. (2023, March 1). *Volkswagen, Audi, and Porsche are getting their own in-car app store—and yes, that includes TikTok.* Retrieved from The Verge: https://www.theverge.

com/2023/3/1/23619468/volkswagen-audi-porsche-cariad-app-store-tiktok.

Guillaume, G. (2022, April 27). *Faurecia-Aptoide signs up Mercedes in infotainment win over Google.* Retrieved from Reuters: https://www.reuters.com/technology/faurecia-aptoide-signs-up-mercedes-infotainment-win-over-google-2022-04-28/.

Haskins, J. (2022, August 23). *What Is The CCPA? Here's How To Comply.* Retrieved from Forbes: https://www.forbes.com/advisor/business/what-is-ccpa/.

JATO Dynamics. (2022). *Affordable EVs and Mass Adoption.* Uxbridge: JATO Dynamics.

NADA. (2022). *NADA Data 2022.* Detroit: NADA.

Nickel Institute. (2021). *Total Cost of Ownership (TCO) for Electric Vehicles (EV) vs Internal Combustion Engine Vehicles (ICE).* Toronto: Nickel Institute.

Schmidt, B. (2021, August 2). *Volkswagen forges new JV to avoid repeat of ID.3 software problems.* Retrieved from The Driven: https://thedriven.io/2021/08/02/volkswagen-forges-new-jv-to-avoid-repeat-of-id-3-software-problems/.

Simoudis, E. (2017). *The Big Data Opportunity In Our Driverless Future.* Corporate Innovators.

Simoudis, E. (2020, June 24). *The Urban Mobility Metric.* Retrieved from Re-Imagining Corporate Innovation with a Silicon Valley Perspective: www.corporateinnovation.co.

Yan, Z., & Goh, B. (2023, May 10). *China's BYD cuts starting price for Seal EV; aims to extend lead.* Retrieved from Reuters: https://www.reuters.com/business/autos-transportation/chinas-byd-launches-lower-priced-versions-seal-ev-aims-extend-lead-2023-05-10/.

Chapter 7

Boudette, N. E. (2022, March 2). *Ford Splits Into Electric and Gas Divisions to Speed Up Transition* . Retrieved from The New York Times: https://www.nytimes.com/2022/03/02/business/economy/ford-model-e.html.

Brock, H. (2022, September 26). *Survey finds 74% of participants prefer to buy EVs at dealerships.* Retrieved from Automotive News: https://www.autonews.com/retail/why-many-ev-buyers-still-prefer-dealerships.

Colias, M. (2023, May 9). *GM Hires Former Apple Cloud Executive to Oversee Software.* Retrieved from The Wall Street Journal: https://www.wsj.com/articles/gm-hires-former-apple-cloud-executive-to-oversee-software-aa8e9b82.

Doll, S. (2021, July 20). *Toyota's Woven Planet acquires Lyft's Level 5 self-driving division.* Retrieved from Electrek: https://electrek.co/2021/07/20/toyotas-woven-planet-acquires-lyfts-level-5-self-driving-division/.

Economist. (2022, November 14). *The race to reinvent the car industry.* Retrieved from The Economist: https://www.economist.com/business/2022/11/14/the-race-to-reinvent-the-car-industry?.

Ferris, D. (2022, September 21). *Hertz places a risky wager on EVs.* Retrieved from Energy Wire: https://www.eenews.net/articles/hertz-places-a-risky-wager-on-evs/.

George, P. (2023, April 4). *Everybody hates GM's decision to kill Apple CarPlay and Android Auto for its EVs.* Retrieved from The Verge: https://www.theverge.com/2023/4/4/23669523/gm-apple-carplay-android-auto-ev-restrict-access.

Gourville, J. T. (2006, June). *Eager Sellers and Stony Buyers: Understanding the Psychology of New-Product Adoption.* Retrieved from Harvard Business Review: https://hbr.org/2006/06/eager-sellers-and-stony-buyers-understanding-the-psychology-of-new-product-adoption.

Hanna, D. (2010, March 8). *How GM Destroyed Its Saturn Success.* Retrieved from Forbes: https://www.forbes.com/2010/03/08/saturn-gm-innovation-leadership-managing-failure.html.

Hearst. (2021, November 6). *Chevrolet MyLink: Everything You Need To Know.* Retrieved from Car and Driver.

Johnson, P. (August, 2022 16). *Ford, GM, BMW partner with SMUD on managed EV charging pilot.* Retrieved from Electrek: https://electrek.co/2022/08/16/ford-gm-bmw-partner-ev-pilot/.

Kacher, G. (2022, November 4). *The secrets of BMW's upcoming Neue Klasse revealed.* Retrieved from Car: https://www.carmagazine.co.uk/car-news/tech/bmw-neue-klasse/.

Kim, C.-Y. (2022, July 29). *Hyundai Motor to buy Korean self-driving startup 42dot.* Retrieved from The Korea Economic Daily: https://www.kedglobal.com/future-mobility/newsView/ked202207290022.

Kitman, J. L. (2022, April 2). *One Weekend in Vegas With the Nation's Auto Dealers.* Retrieved from The New York Times: https://www.nytimes.com/2022/04/02/business/auto-dealers-sales-profits.html.

Lutz, Z. (2011, October 12). *Cadillac unveils CUE infotainment system for connected driving excitement in 2012.* Retrieved from Engadget:

https://www.engadget.com/2011-10-12-cadillac-unveils-cue-infotainment-system-for-connected-driving-e.html.

Martinez, M. (2023, March 15). *29 Ford dealers drop out of EV program after changes.* Retrieved from Automotive News: https://www.autonews.com/dealers/29-ford-dealers-leave-ev-program-after-changes.

Meiners, J., and Knoedler, M. (2013, May 22). *A Glimpse Inside BMW's i Electric Project.* Retrieved from Car and Driver: https://www.caranddriver.com/news/a15372489/a-glimpse-inside-bmws-i-electric-project/.

Mihalascu, D. (2022, October 5). *Sixt To Purchase 100,000 BYD EVs For Its European Fleet By 2028.* Retrieved from Inside EVs: https://insideevs.com/news/614438/sixt-purchase-100000-byd-evs-for-its-european-fleet-by-2028/.

Paul, G. (2023, February 15). *MERCEDES-BENZ NAMES A SOFTWARE CHIEF TO STEER ITS DIGITAL FUTURE.* Retrieved from Theorg: https://theorg.com/iterate/mercedes-benz-names-a-software-chief-to-steer-its-digital-future.

Posky, M. (2022, May 23). *Mercedes Ending Dealer Sales Model in Europe.* Retrieved from The Truth About Cars: https://www.thetruthaboutcars.com/2022/05/mercedes-ending-dealer-sales-model-in-europe/.

PwC. (2022). *PwC's 2022 Car Consumer and Dealer Survey.* PwC.

Ravi, Anjan. (2022, December 27). *7 benefits of the Mercedes MMA platform in the company's words.* Retrieved from TopElectricSUV.com: https://topelectricsuv.com/news/mercedes-benz/2025-mercedes-eqc-details/.

Reed, A., MacDuffee, J. P., and Hrebiniak, L. (2021). *Saturn: A Wealth of Lessons from Failure.* Retrieved from Knowledge at

Warton: https://knowledge.wharton.upenn.edu/article/saturn-a-wealth-of-lessons-from-failure/.

Simoudis, E. (2020). *Transportation Transformation.* Corporate Innovators.

Simoudis, E. (2021, August 2). *Will Automakers Become Apple, AT&T, or Foxconn?* Retrieved from Re-Imagining Corporate Innovation: https://corporate-innovation.co/2021/08/02/will-automakers-become-apple-att-or-foxconn/.

Simoudis, E. (2023, June 1). *Ford CMD: The Right Goals Are Set; But Can The Company Execute?* Retrieved from Re-Imagining Corporate Innovation: https://corporate-innovation.co/2023/06/01/ford-cmd-the-right-goals-are-set-but-can-the-company-execute/.

Staff, N. (2009, April 3). *How GM Crushed Saturn.* Retrieved from Newsweek: https://www.newsweek.com/how-gm-crushed-saturn-77093.

Torsoli, A. (2022, November 7). *Renault Seeks Margin Boost From Sweeping Overhaul Plan.* Retrieved from Bloomberg: https://www.bloomberg.com/news/articles/2022-11-08/renault-to-boost-returns-in-split-plan-with-outside-investors?.

Trop, J. (2022, June 2). *Ford wants to restructure its dealership model to boost EV sales.* Retrieved from Techcrunch: https://techcrunch.com/2022/06/02/ford-wants-to-sell-evs-online-only-and-at-a-set-price/.

Wayland, M. (2022, March 2). *Ford will split EVs and legacy autos into separate units as it spends $50 billion on electric vehicles.* Retrieved from CNBC: https://www.cnbc.com/2022/03/02/ford-splits-evs-and-legacy-autos-into-separate-units.html.

Acknowledgments

This book completes a trilogy about the role of AI in new mobility. When I first started writing about the role of AI in driverless mobility, few people were focusing on AI, and new mobility was not yet a well-formulated concept. In the intervening years, a lot has changed both about mobility and more recently about AI's contributions in this field. I have benefited from the help that many individuals have provided me during this period.

Paul Lienert, Stephen Zoeph, and Goetz Weber have been constants during this journey, and their help during the writing of this book was no exception. They steered me in the right direction when my arguments were going off a cliff and encouraged me when I was hitting the proverbial wall. I have benefited greatly from conversations with Nikos Michalakis. My realization that a new customer experience is required as the automotive industry transforms and adopts Software-Defined Vehicles was influenced by these conversations. Subsequent conversations about the methodologies that the implementation of Software-Defined Vehicles will require were also particularly helpful.

My editor Meghan Houser worked hard to make the right ideas come forward and the manuscript better. She helped me effectively connect ideas I first expressed in my previous two books with the narrative of this book thus completing the trilogy. Ad'm DiBiaso did a great job in generating wonderful graphics out of my scribbles and creating the book's cover.

Numerous other people from industry, government, and academia provided me with ideas, read the manuscript, and offered extremely valuable feedback. I want to acknowledge Joe White for all the comments and suggestions he provided throughout this process, and for his friendship. I also want to thank the industry executives who offered me their time during my research.

As with the previous two books, many of the ideas presented in this book benefited from extensive discussions I had with the members of our firm's advisory board. Our conversations made the book's content richer and helped me focus on the specific issues the automotive and AI industries will have to address.

Finally, I'd like to thank my wife, Kathleen, for all her continuous encouragement and patience throughout this project. Even though this is my third book, not unlike the previous two, it wouldn't have been possible without her amazing contributions, small and large.

About the Author

Evangelos Simoudis is a recognized expert on AI, big data, and new mobility. He has been working in Silicon Valley for thirty years as a startup founder and CEO, corporate executive, venture investor, and senior advisor to global corporations and governments. His firm, Synapse Partners, is investing in AI startups and advising senior management teams of large organizations on AI and new business models. In addition to being presented in several papers, the results of his work on new mobility have been published in two books: *The Big Data Opportunity in Our Driverless Future* and *Transportation Transformation.* He is a member of the California Institute of Technology's (Caltech) Advisory Board, the Advisory Board of Brandeis International Business School, the Advisory Board of C2SMART (a US Department of Transportation Center of Excellence), and the Advisory Board of Securing America's Future Energy. He earned a PhD in computer science from Brandeis University in machine learning and large databases and a BS in electrical engineering from Caltech. Connect with Evangelos at evangelossimoudis.com and on LinkedIn at linkedin.com/in/evangelossimoudis.

INDEX

Note: Page numbers in italic refer to figures. Page numbers followed by "n" with numbers refer to footnotes.

A

Abbott, Michael, 161

ADAS. *See* Advanced Driver Assistance System

Advanced Driver Assistance System (ADAS), 68, 73, 78, 88, 102
  functionality in BlueCruise, 160
  technologies, 21–22

Afeela, 81, 112, 161–163

agile/agility, 9, 176
  of automakers, 9
  operational, 186
  product development processes, 96
  software development process, 90, 96, 99, 159, 171

AI. *See* artificial intelligence

Amazon, 8, 49, 97
  Amazon Prime, 9

customer relationship, 156, 182
in-cabin technologies, 61

AMG.EA platform, 158

Android Studio (Google), 105

Apache Eclipse IDE, 107

APIs. *See* application programming interface

Apple, 8, 49, 97
controlling consumer's digital life, 182
customer relationship, 156
in-cabin technologies, 61

application programming interface (APIs), 70, 94–95, 96, 106, 108

applications
AI, 61–62
CRM, 120
customer-facing, 21
mobile, 105–106
for vehicle's operation, 105

artificial intelligence (AI), 23, 34, 37, 84, 103
AI-based digital platforms, 21
component in Vehicle Management Platform, 116
FEAT framework, 124–126
generative AI system, 65–66
models version control system, 103–105
in Software-Defined Vehicles, 183–184

assembly line
car production, 38n3
invention, 1n1

Audi, 103

automakers, 190
assembly line car production, 38n3
choice of business model, 9
using clean sheet approach, 80
as customer's mobility partner, 49

goal for new mobility user requirements, 26
innovations, 2–3, 139–140
Microcontroller-Based Architectures, 67–68, 71
need to remake partner ecosystem, 9–10
on-demand mobility services, 21
optimizations of vehicle's performance, 83–84
producing ICE engines, 139
responsibility to customer demand for vehicles, 2
transformation, 8, 10–11, 139

automotive customer experience, 30
of BMW i-brand vehicles, 149–150, 154
flagship vehicle, 38, 38n4
personalizable, 189–190
redesign, 39–40
variation by vehicle class and model, 45–46

automotive customer journey, 29, 30, 42–43
end-to-end, 31–32, 32, 188

automotive industry, 1–6, 170–172, 179

AutoNation, 120

AWS, 81, 157

Azure, 157

B

Barra, Mary, 109

battery cost, 10, 65

battery technology lifecycle, 98

Bill of Materials cost (BOM cost), 65–66

BlueCruise, 160

BMW, 60, 67

BMW Project i brand vehicles, 144, 147
cultures, 150–151, 171

customer experience, 149–150
innovations, 148–149, 154
BOM cost. *See* Bill of Materials cost
Bosch, 67, 186
brand-defining, 57, 124
brand ladder, 38
Brightdrop (GM), 19
BYD, 130, 163, 170, 181

C

Cadillac, 38, 41, 78, 164
Cariad, 140
Central Computer Architectures, 69, 71, 72, 76–77, 80, 81, 93, 185
Chesnut, Donald, 161, 162, 168
Chinese OEMs, 8, 130, 163
CI/CD. *See* Continuous Integration/Continuous Delivery
clean sheet Software-Defined Vehicle, 77, 78–80, 101, 180
Compact Crossover Utility Vehicle (CCUV), 52–56
Continuous Integration/Continuous Delivery (CI/CD), 90, 171–172
convenience for customers, 18, 20, 35, 42, 56, 59, 64, 169
cradle-to-grave vehicle journey, 31–32, 35–37, 36, 162, 163
CRM applications. *See* customer relationship management applications
customer-centricity, 30–31, 49, 117, 119, 165, 171–173, 188
Customer Data Repository, 117, 118, 120
customer experience. *See* automotive customer experience
customer journey. *See* automotive customer journey
Customer Management Platform, 82, 111, 116–120, 118

customer profile, 118–119
creation in FEAT framework, 122–123
generic, 188

customer relationship management applications (CRM applications), 120

customer(s)
lifetime value, 3, 40, 46–47, 83–84, 129, 132, 179, 186
mobility-related activity, 33–35

cybersecurity services, 58, 94, 115, 119

D

dealers, 10, 29, 153–154, 167–169

DevOps, 90, 99

digital twins, 98, 100–101

Domain-Based Software-Defined Vehicles, 71, 72, 72–74, 93, 110, 111, 155–156, 180, 182, 187

Durach, Stephan, 143, 163–164

E

ECUs. *See* Electronic Control Units

E/E architectures, 70, 78, 79

electric vehicles, 2–4, 6–7, 10, 51, 65, 69, 78, 127, 130–131, 148, 155, 160, 166, 189. *See also* Software-Defined Electric Vehicles

Electronic Control Units (ECUs), 67, 72, 95

end-to-end customer journey, 7, 31–32, 33, 188, 190

F

factory automation, 17–18

FEAT framework. *See* Flagship Experience AI Technology framework

Flagship Experience, 7, 9, 11–12, 19–20, 23, 27, 40, 45, 52–56, 133, 142, 180
adoption and monetization, 126–132
for Business, 43–44, 45
cradle-to-grave vehicle journey, 31–32, 35–37, *36*, 162
culture clash, 171–174
customer-centricity, 30–31, 188
Customer Management Platform, 111, 116–120
Data and AI, 183–184
end-to-end customer journey, 31–32, *33*
FEAT framework, 121–126, 189
features and services, 56–62
measuring customer's satisfaction level, 133–134
mobility-related activity, 33–35
mobility-related relationship between customer and OEM, 44
positioning automaker as customer's mobility partner, 49
pre-sale process, 123
radical transformation, 10–11
scenarios in customer's relationship, 41–43
SDV Services Marketplace, 111–113
Software-Defined Vehicles, 7–8, 78, 81–85
software relating to, 88–89, *89*
subscription-based business models, 132–133
Tesla's approach to customer experience, 49–52
transformations, 143, 162–166
US and EU investments, 182–183
Vehicle Management Platform, 111, 113–116
Zone-Based Architectures, 75

Flagship Experience AI Technology framework (FEAT framework), 121–126, 189

AI corporate maturity levels, 125–126
AI models, 124–125
customer profile creation, 122–124
data types, 121–122
Urban Mobility Metric, 122

flagship vehicle, 38, 38n4

fleet formation phase, 22–24

Ford, 52, 66, 84
defining dealer's role, 167–168
FordPro, 19
Mustang Mach-E vehicles, 22, 57, 73, 78
platforms, 73, 78
transformation, 141, 159–161

G

General Motors (GM), 10, 41, 48, 52, 60, 66, 84, 103
clean sheet software-defined vehicles, 78, 79
Cruise robotaxi unit, 25–26
defining dealer's role, 167–168
monetization of personal services, 127
organizational transformation goals, 161
revenue for warranty costs, 66
revenue from services, 109–110
Saturn Corporation, 144–147, 152
transformation, 141, 161

GM. *See* General Motors

goods delivery, 2, 18–19, 21, 23

Google, 8, 49, 97, 103
customer relationship, 156, 182
in-cabin services, 60, 61, 166

GPT-4 (OpenAI), 66

H

hardware platform, 69, 70, 77–78, 113
Hester, Travis, 45, 54, 61
High-Performance Computing (HPC), 68–69
Houston, Jonah, 39, 44, 61
HPC. *See* High-Performance Computing
Hyundai
    Software-Defined Vehicles, 93, 157

I

ICE vehicles. *See* Internal Combustion Engine vehicles
IDEs. *See* Integrated Development Environments
Integrated Development Environments (IDEs), 105–108
intelligent simulator, 101–103
Internal Combustion Engine vehicles (ICE vehicles), 4–5, 11, 139, 141, 182

L

large-scale manufacturing, 180–181

M

machine learning, 75, 97, 103, 104, 107, 108
Mercedes, 60, 67, 84
    transformation, 93, 127, 141, 167, 176, 181
    platforms, 157

Microcontroller-Based Architectures, 67–68, 71, 72, 77–78, 85

middleware, 91, 93–94

mobility, 15–16

monetizable/monetization, 3
    capabilities, features, and services, 57, 60–61
    of Flagship Experience, 12, 126–135
    opportunities, 61, 84
    in Software-Defined Vehicle, 36, 47, 182

N

new mobility, 1–2, 11, 19, 190. *See also* Flagship Experience
    changes in patterns, 16–19
    fleet formation phase, 22–24
    mobility-as-a-service phase, 24–25
    mobility services emergence phase, 20–22
    OEMs, 3, 19–20
    user requirements, 26–27

NVIDIA, 74, 81, 103

O

OEM Cloud, 70, 82, 94, 98, 108, 111

OEMs. *See* Original Equipment Manufacturers

OEM transformation effort, 143–144. *See also* Original Equipment Manufacturers (OEMs)
    BMW Project i, 144, 147–150
    dealers role, 153–154
    GM's Saturn Corporation, 144–147
    importance of innovations, 154–155
    organizational culture, 150–152

organizational and business transformations, 12, 137–138
to create and offer Flagship Experience, 162–166
defining dealer's role value chain, 167–169
hiring eligible employees, 174–176
OEM categories, 138–143
past OEM transformation effort, 143–155
preparation for culture clash, 171–174
resolving conflicts in value chain, 169–171
along technology dimension, 155–162

Original Equipment Manufacturers (OEMs), 3, 5–6, 31, 32, 39, 63, 134
AI system, 34, 37
competitors, 43–44
customer-centricity, 179, 180, 181–182
customer's mobility-related activity, 33, 40–41, 44
digital twin, 100–101
Domain-Based Architectures, 182, 187
end-to-end customer journey, 32, *33*
features and services of Flagship Experience, 56–62
using Flagship Experience, 44–45
hiring eligible employees, 175–176
increasing customer lifetime value, 46–47
large-scale manufacturing, 180–181
monetizing Software-Defined Vehicles, 131–132, 184–185
Parc, 7, 7n2
partnership, 9–10, 48, 49, 170
responsibility, 173–174
scenarios in customer's relationship, 41–43
Software-Defined Vehicles, 19–20, 71, 93, 139, 141–142, 182, 187
technology and business transformation, 139–140
transformations, 176–177
vehicle-centricity, 29–30, 180

Ostberg, Magnus, 157

OTA software updates. *See* over-the-air software updates

over-the-air software updates (OTA software updates), 22, 63–64, 98–99

P

Parc, 7, 7n2, 71, 163

personal features or services, 57–58, 124

R

radical transformation, 10–11

real-time operating system, 91, 93, 107

Renault, 60
  operating system, 156
  organizational transformation, 159–160
  technology-focused company, 158
  technology stacks, 185

retrofit approach, 77–78, 187

Rivian, 39
  software tooling platforms, 97

S

Saturn Corporation of GM, 144–145, 150
  business model innovation, 145–146
  corporate innovations, 145
  organizational culture, 150–151, 171
  technical innovations, 146, 154

SDV Services Marketplace, 111–113, 119, 124

service ideation, 163–164

service-oriented architecture (SOA), 96

skateboard, 70, 96

SOA. *See* service-oriented architecture

Software-Defined Vehicles, 2, 8, 12, 19–20, 22, 63, 190
ADAS technologies, 21
adoption of, 134
AI applications, 61–62
architectures, 68, *70*, 71
capabilities, features, and services, 56–57
Central Computer Architectures, 71, 72, 76–77
clean sheet approach, 77, 78–80
cradle-to-grave vehicle journey, 35–37, 36, 163
customer monetization in, 182
development, 141–142
Domain-Based Architectures, 71, 72, 72–74
extensibility and upgradability, 84–85
Flagship Experience, 7–8, 46, 78, 85
hardware platform, 69, 70
hardware-related BOM cost, 65–66
impact on Flagship Experience, 81–84
intelligence of, 26–27
model in OEM's lineup, 45–46
OTA software updates, 63–64
principles, 68–69
retrofit approach, 77–78
skateboard, 69, 70
software platform, 69
technology development, 139, 185–186
US and EU investments, 182–183
using Zone-Based Architectures, 74–75
Zone-Based Architectures, 71, 72, 74–76

software for Software-Defined Vehicle, 87

software development process, 89–91

Software Platform, 88–89, 91–96

Software Tooling, 88–89, 97–108

software lifecycle, 98

Software Platform, 88–89, 91, 113
upgrading vehicle's components, 94–95

Software Tooling, 88–89, 97–98
AI models version control, 103–105
digital twins, 100–101
IDEs, 105–108
intelligent simulator, 101–103

Stellantis, 67, 156, 174, 176, 185

subscription-based business models, 132–133

T

TCO. *See* Total Cost of Ownership

technology, transformations, 155–158
in OEMs, 159–162

Tesla, 10, 84
approach to customer experience, 49–52
customer-centricity, 29–30, 123
focus on Software-Defined Vehicles' architecture, 74, 80n7, 93, 97
monetization of personal services, 127
OTA software, 64

Total Cost of Ownership (TCO), 56, 130–131

Toyota, 52, 60, 67, 103

transportation, 15–16. *See also* new mobility
impact on climate change, 186

transformation approaches, 25–26, 138

U

Uber, 141
  customer-facing application, 21

Urban Mobility Metric, 122

V

value chain, 167–171

vehicle-centric(ity), 29–30, 39, 180
  relationship with owners, 47–48

Vehicle Application IDE, 106–107

vehicle journey
  cradle-to-grave, 31–32, 35–37, *36*, 162, 163
  owners of Software-Defined Vehicle, 47

Vehicle Management Platform, 82, 111, 113, 114, 116. *See also* Customer Management Platform
  OTA update tool, 114–116

Vehicle Model, 100–101

version control system, 103–105

Vision Zero, 179

Volkswagen (VW), 52, 60, 66
  hardware platforms, 73
  software platforms, 88, 93

Volvo, 88, 103, 156

VW. *See* Volkswagen

Z

Zone-Based Architectures, 71, 72, 74–76, 182, 185, 187
  of Software-Defined Vehicle, 81–82, 110–111, 126, 128, 129, 156, 180

zone controllers, 74, 76, 83